ECOLOGICAL ASSESSMENT OF
POLYMERS

STRATEGIES FOR PRODUCT STEWARDSHIP AND REGULATORY PROGRAMS

ECOLOGICAL ASSESSMENT OF
POLYMERS

STRATEGIES FOR PRODUCT STEWARDSHIP AND REGULATORY PROGRAMS

Edited by

JOHN D. HAMILTON and ROGER SUTCLIFFE

VAN NOSTRAND REINHOLD

I(T)P® A Division of International Thomson Publishing Inc.

New York • Albany • Bonn • Boston • Detroit • London • Madrid • Melbourne
Mexico City • Paris • San Francisco • Singapore • Tokyo • Toronto

Copyright © 1997 by Van Nostrand Reinhold

 Van Nostrand Reinhold is an International Thomson Publishing Company
The ITP logo is a registered trademark used herein under license

Printed in the United States of America

For more information, contact:

Van Nostrand Reinhold
115 Fifth Avenue
New York, NY 10003

Chapman & Hall GmbH
Pappelallee 3
69469 Weinheim
Germany

Chapman & Hall
2-6 Boundary Row
London
SE1 8HN
United Kingdom

International Thomson Publishing Asia
221 Henderson Road #05-10
Henderson Building
Singapore 0315

Thomas Nelson Australia
102 Dodds Street
South Melbourne, 3205
Victoria, Australia

International Thomson Publishing Japan
Hirakawacho Kyowa Building, 3F
2-2-1 Hirakawacho
Chiyoda-ku, 102 Tokyo
Japan

Nelson Canada
1120 Birchmount Road
Scarborough, Ontario
Canada M1K 5G4

International Thomson Editores
Seneca 53
Col. Polanco
11560 Mexico D.F. Mexico

1 2 3 4 5 6 7 8 9 10   BKR   01 00 99 98 97 96

Library of Congress Cataloging-in-Publication Data

Ecological assessment of polymers : strategies for product stewardship
    and regulatory programs / editors, John D. Hamilton, Roger
    Sutcliffe——1st ed.
        p. cm.
    Includes bibliographic references.
    ISBN 0-442-02328-6
    1. Ecological risk assessment. 2. Pollution——Environmental
aspects. 3. Environmental policy. I. Hamilton, John D. (John
Douglas), 1960- . II. Sutcliffe, Roger.
QH541 . 15 . R57E24   1996
363 . 738 ' 4——dc21   96—47634
                                                        CIP

http://www.vnr.com
product discounts • free email newsletters
software demos • online resources

email: info@vnr.com

A service of I(T)P®

Contents

Preface

This book will assist readers in identifying improvements in ecological test methods with polymers, in reviewing the performance of degradable polymers, and in developing credible regulatory control and environmental labeling programs. No other current overview exists in the general toxicology literature regarding ecological assessment of commercial synthetic polymers. We have included environmental regulatory assessment schemes that are designed to prevent unreasonable effects of polymers on plant, acquatic, and terrestrial organisms. In addition, the book provides author interpretations of ecotoxicity and fate tests with polymers as well as an overview of new directions in degradable polymer research and Life-Cycle Assessment.

We gratefully acknowledge the following people who helped to ensure the development and success of this book: Rafael Aviles, Nancy Bordage, Linda Burg, Kenneth Dickson, Patric Friend, Wladyslaw Grodzinski, Isadore Morici, Shaunalea Savard, Harvey Scribner, David Shelton, Melonie Wahl, and Ernie Weiler. The editors and authors are also grateful to Rohm and Haas Company and S.C. Johnson Wax for providing the essential overall support for this book.

Contributors

Andrew J. Atkinson
Environment Canada
Hull, Quebec, Canada K1A 0H3

Richard Bartha, Ph.D.
Rutgers University
New Brunswick, NJ, USA 08903

Robert Boethling, Ph.D.
U.S. EPA (MS 7406)
Washington, DC 20460

Michael A. Cole, Ph.D.
University of Illinois
Urbana-Champaign, IL,
USA 61801

Sergio S. Cutié, Ph.D.
Dow Chemical Company
Midland, MI, USA 48674

Jon Doi, Ph.D.
Roy F. Weston, Inc.
Lionville, PA, USA 19341

Michael B. Freeman, Ph.D.
Rohm and Haas Company
Bristol, PA, USA 19007

Stanley J. Gonsior, Ph.D.
Dow Chemical Company
Midland, MI, USA 48674

Patrick D. Guiney, Ph.D.
S.C. Johnson Wax
Racine, WI, USA 53403

John D. Hamilton, Ph.D., DABT
S.C. Johnson Wax
Racine, WI, USA 53403

Krzysztof M. Jop, Ph.D.
Science Applications
International Corp.
Narragansett, RI, USA 02882

Larry A. Lyons
Betz Laboratories, Inc.
Trevose, PA, USA 19047

Ronald L. Keener, Ph.D.
Rohm and Haas Company
Philadelphia, PA, USA 19105

Robert A. Matheson
Environment Canada
Hull, Quebec, Canada K1A 0H3

Vincent McGinniss, Ph.D.
Battelle–Columbus
Columbus, OH, USA 43201

James E. McLaughlin, Ph.D.
Rohm and Haas Company
Spring House, PA, USA 19477

Isadore J. Morici
Rohm and Haas Company
Spring House, PA, USA 19477

J. Vincent Nabholz, Ph.D.
U.S. EPA (MS 7403)
Washington, DC, USA 20460

Marie-Pierre Rabaud, Ph.D.
Exxon Chemical France
Cedex 31–92098, Paris, France

Gordon Reidy, Ph.D.
Woodward-Clyde Pty Limited
St. Leonards, NSW, Australia 2065

Kevin H. Reinert, Ph.D.
Rohm and Haas Company
Spring House, PA, USA 19477

Graham Swift, Ph.D.
Rohm and Haas Company
Spring House, PA, USA 19477

Duane A. Tolle, Ph.D.
Battelle–Columbus
Columbus, OH, USA 43201

Stephen R. Vasconcellos, Ph.D.
Betz Laboratories, Inc.
Trevose, PA, USA 19047

Bruce W. Vigon, Ph.D.
Battelle–Columbus
Columbus, OH, USA 43201

Asha V. Yabannavar, Ph.D.
Rutgers University
New Brunswick, NJ, USA 08903

ECOLOGICAL ASSESSMENT OF
POLYMERS

STRATEGIES FOR PRODUCT STEWARDSHIP AND REGULATORY PROGRAMS

PART *1* | *Introduction and overview*

Chapter 1

Introduction

■ JOHN D. HAMILTON,[a] STEPHEN R. VASCONCELLOS,[b]
AND RONALD L. KEENER[c]
S. C. Johnson Wax,[a] BetzDearborn, Inc.,[b] and Rohm and Haas Company[c]

SCOPE

Products that contain synthetic polymers are used in virtually every conceivable technology, including agriculture, biomedicine, cleaning, clothing, construction, disinfection, electronics, engineering, food processing, mining, transportation, and wastewater treatment. This book focuses on synthetic polymers that may be discharged to the environment. Wastewater containing water-soluble polymer dispersants used in detergents can reach treatment facilities that discharge to natural aquatic environments. Other polymers are added directly to water during wastewater treatment operations. Polymers that are found in coating products such as paint may reach wastewater from polymer manufacturing as well as during product formulation and use. Further, there is considerable interest in the use of degradable plastics in waste management programs. The following chapters provide ecological regulatory assessments of polymers, author interpretations of ecotoxicity and fate tests with polymers, and overviews of degradable polymer research and Life-Cycle Assessment (LCA).

Ecotoxicology test data for polymers are based on studies of aquatic organisms, terrestrial invertebrates, and vascular plants. Toxicity tests with mammalian or avian test organisms tend to be outside the scope of existing ecotoxicology test programs with polymers, and reviews on the mammalian toxicity of polymers are subjects of mammalian toxicology and human health risk assessment. However, it should be noted that commercial synthetic polymers are consistently of low inherent toxicity and irritation potential to mammals, with the greatest potential for respiratory toxicity and/or irritation being associated with polymer dust exposure (e.g., CEFIC, 1995). The nonspecific nature of inhalation toxicity that has been reported with polymers is consistent with respiratory clearance overload associated with high levels of exposure to biologically inert particulates. The respiratory effects do not appear to be associated with low-dose exposure. There is a wide array of additives to polymeric products (e.g., metals, surfactants, and organic solvents), and the reader is referred to health overviews on additive chemicals (e.g., Cheremisinoff, 1995).

The high molecular weight and large steric size of synthetic polymers substantially limit absorption of polymers across biological membranes. However, the chemical reactivity of polymer functional groups can influence polymer ecotoxicity and/or ecological fate. Also, many polymers consist of a distribution (i.e., range) of molecular weight, not a single molecular weight. Molecular weight distribution can alter the degree of biodegradability and sorption (i.e., partitioning) of some polymers. The influence of solubility in both laboratory and natural environments also must be considered in assessing the inherent ecotoxicity and ecological fate of polymers. The effects of molecular weight, solubility and functional groups are discussed in this book.

Regulatory authorities worldwide are developing polymer-specific criteria for ecological assessment. The Organisation for Economic Co-operation and Development (OECD) has developed guidelines on solution-extraction behavior of polymers in water as well as on the determination of the low molecular weight content, number-average molecular weight, and molecular weight distribution of polymers using gel permeation chromatography (OECD, 1995a–c). The European Union (E.U.) has written a guidance document for implementation of Council Directive 67/548/EEC, published as Commission Directive 93/105/EEC, on the tests required in the E.U. when polymers are required to be notified as new chemical substances (E.U., 1995). Industrial trade organizations have provided the E.U. and OECD with background information on polymer characteristics that can influence ecotoxicity (e.g., CEFIC, 1995) to demonstrate that many polymers can be classified as "low concern" for ecological toxicity. The United States, Canada, and countries in the Pacific region also consider the physical-chemical properties of polymers (i.e., molecular weight, solubility, and functional groups) in regulating polymers. Examples of the influence of these properties are provided in Part 2. Regulatory approaches to the ecological assessment of polymers are outlined in Part 3.

New directions in polymer research and development are driven by market demands. Industry has responded to an interest in environmentally biodegradable polymers by initiating specific synthesis research programs to design environmentally degradable polymers. Assessment of the overall ecological significance of any new biodegradable polymer design should consider raw material resource expenditure, disposal attributes, and relative performance compared to existing products. Life-Cycle Assessment (LCA) is an ecological assessment approach that facilitates the comparison. Part 4 provides an overview of environmentally degradable polymer research and an outline of LCA as it applies to polymers.

This chapter introduces ecological assessment and synthetic polymer chemistry associated with polycarboxylates, polyacrylate superabsorbents (PAS), water treatment polymers, emulsions and related polymers, and plastics. Further details can be found in specific texts, including those written on ecological risk assessment (e.g., Suter, 1993) and polymerization chemistry (e.g., Odian, 1991). The description of polymer chemistry used here is based on prior overviews (NPCA, 1981; Odian, 1991).

ECOLOGICAL ASSESSMENT

Ecological risk assessment is a specific form of ecological assessment, where risk assessment principles are used to evaluate the probability and/or the magnitude of significant changes in an ecosystem from physical, chemical, and/or biological stresses. More broadly, ecological assessment includes ecotoxicity and ecological fate testing, regulatory reviews, degradable polymer evaluations, and LCA.

Part 3 provides regulatory criteria for ecological assessment of polymers. However, it is difficult to understand the significance of test results with regulatory criteria alone. Tables 1-1 and 1-2 provide generic (i.e., nonregulatory) criteria to communicate the results of ecotoxicity and ecological fate studies.

COMMERCIAL SYNTHETIC POLYMERS

General Aspects

This book focuses on commercial synthetic polymers. These polymers are found in a broad range of molecular weights from about 1,000 to over 1 million, and are generally synthesized from organic polymerization reactions. An important exception to the production of commercial synthetic

Table 1.1. Generic communication criteria for ecotoxicity[a]

Characteristic	Description	Measurement[b]
Acute aquatic toxicity	High concern	$EC_{50} \leqslant 1$ mg/L
	Moderate concern	$EC_{50} > 1$ mg/L and $\leqslant 100$ mg/L
	Low concern	$EC_{50} > 100$ mg/L
Acute sediment toxicity	High concern	$EC_{50} \leqslant 1$ mg/kg dry weight sediment
	Moderate concern	$EC_{50} > 1$ mg/kg and $\leqslant 100$ mg/kg
	Low concern	$EC_{50} > 100$ mg/kg
Acute toxicity: terrestrial wildlife	High concern	$LD_{50} \leqslant 100$ mg/kg body weight
	Moderate concern	$LD_{50} > 100$ mg/kg and $\leqslant 2{,}000$ mg/kg
	Low concern	$LD_{50} > 2{,}000$ mg/kg
Acute toxicity: terrestrial wildlife exposed in diet	High concern	$LC_{50} \leqslant 1{,}000$ mg/kg dry weight diet
	Moderate concern	$LC_{50} > 1{,}000$ mg/kg and $\leqslant 5{,}000$ mg/kg
	Low concern	$LC_{50} > 5{,}000$ mg/kg
Acute toxicity: plants and soil organisms	High concern	$EC_{50} \leqslant 1$ mg/kg dry weight soil
	Moderate concern	$EC_{50} > 1$ mg/kg and $\leqslant 100$ mg/kg
	Low concern	$EC_{50} > 100$ mg/kg
Chronic aquatic toxicity	High concern	Chronic value $\leqslant 0.1$ mg/L
	Moderate concern	Chronic value > 0.1 mg/L and $\leqslant 10$ mg/L
	Low concern	Chronic value > 10 mg/L
Chronic sediment toxicity	High concern	Chronic value $\leqslant 1$ mg/kg dry weight sediment
	Moderate concern	Chronic value > 1 mg/kg and $\leqslant 10$ mg/kg
	Low concern	Chronic value > 10 mg/kg
Chronic toxicity: terrestrial wildlife exposed in diet	High concern	$LD_{50} \leqslant 100$ mg/kg dry weight food
	Moderate concern	$LD_{50} > 100$ mg/kg and $\leqslant 500$ mg/kg
	Low concern	$LD_{50} > 500$ mg/kg
Chronic toxicity: plants and soil organisms	High concern	$EC_{50} \leqslant 0.1$ mg/kg dry weight soil
	Moderate concern	$EC_{50} > 0.1$ mg/kg and $\leqslant 10$ mg/kg
	Low concern	$EC_{50} > 10$ mg/kg

[a]Adapted from U.S. EPA, 1992 and Smrchek et al., 1993.
[b]EC_{50} — exposure concentration causing an adverse effect (e.g., lethality) in 50% of the study population; LD_{50} — dose causing lethality in 50% of the study population; Chronic value — no observed (adverse) effect concentration (i.e., NOEC or NOAEC) or Maximum Allowable Toxicant Concentration (MATC): MATC = estimated from the geometric mean of the NOEC or NOAEC and the LOEC; LOEC = the lowest concentration that causes an effect that is statistically different from the controls.

polymers by organic polymerization reactions is the use of biotechnology to produce environmentally biodegradable polyester plastics.

The word polymer originates from the Greek *poly* (many) and *meros* (parts or units). A polymer chain (i.e., backbone) is a group of many

Table 1.2. Generic communication criteria for ecological fate[a]

Characteristic	Description	Measurement
Water solubility	Very soluble	$> 10,000$ mg/L
	Soluble	$> 1,000$ to $10,000$ mg/L
	Moderately soluble	> 100 to $1,000$ mg/L
	Slightly soluble	> 0.1 to 100 mg/L
	Insoluble	< 0.1 mg/L
Organic partition coefficient	Very strong sorption	$\geqslant 4.5$
	Strong sorption	$\geqslant 3.5$ and < 4.5
	Moderate sorption	$\geqslant 2.5$ and < 3.5
	Low sorption	$\geqslant 1.5$ and < 2.5
	Negligible sorption	< 1.5
Biodegradation	Rapid	$\geqslant 60\%$ degradation[b] over a test period $\leqslant 7$ days
	Moderate	$\geqslant 30\%$ degradation over a test period $\leqslant 28$ days
	Slow	$< 30\%$ degradation over a test period $\leqslant 28$ days
	Very slow	$< 30\%$ degradation over a test period > 28 days

[a]Adapted from U.S. EPA, 1992.
[b]Degradation = mineralization measured by CO_2 evolution, loss of dissolved organic carbon (DOC), or biochemical oxygen demand (BOD).

structural units, typically called chemical reactants or monomers. Regulatory agencies can differ considerably in terms of assessment and definition of polymers. Broadly defined, however, a typical synthetic commercial polymer is a substance that consists of multiple covalently bound monomers and/or other chemical reactants. Polymers are identified (i.e., characterized) by the type and the sequence of the monomers. For example, acrylic polymers based on acrylic acid and maleic acid can be generically identified as polyacrylic/maleic acid. A plastic is generally a high molecular weight polymer; "plastic" usually refers to the way a solid polymer will melt and flow under thermal stress.

Those involved in the ecological assessment of synthetic polymers should understand the basic terminology of polymer chemistry. The systematic nomenclature of polymers is normally based on the Chemical Abstract Service Registry Number (CASRN) system or the International Union of Pure and Applied Chemistry (IUPAC) system. Other common systems of polymer classification are based on (1) the type of reaction mechanism during polymerization (e.g., chain and step reactions), (2) polymer synthesis (e.g., addition and condensation methods), (3) the source of the primary monomer (e.g., polyethylene from ethylene, polyvinyl chloride from vinyl chloride, etc.), or (4) the degree of branching (linear, branched, and cross-linked polymer backbone chains).

Both the CASRN and the IUPAC systems provide precise structure-based definitions of polymers. However, the CASRN system has several advantages over the IUPAC system, as (1) CASRN assigns specific numerical identifiers to unique chemical structures, and (2) CASRN is better able to deal with complex natural substances and with substances of unknown or variable composition. Consequently, CASRN is the more appropriate and preferred method for regulatory polymer-specific nomenclature. However, broader forms of classification such as the primary monomer source method are often sufficient to cluster polymers into categories of predicted ecotoxicity and fate characteristics. Use of the polymer synthesis method to classify polymers also assists in ecological assessments. The high molecular weight and insolubility in water of many nonreactive polymers produced by addition polymerization, for example, will significantly limit the potential for ecotoxicity. The ability of some polymers created by condensation polymerization to biodegrade can be a favorable environmental attribute. The CASRN and the IUPAC systems occasionally provide systematic polymer names that are not commonly known in the field of ecological assessment (e.g., polystyrene from the common source method is poly(1-phenylethylene) in the IUPAC system). Therefore, for the specific purposes of this book, primary monomer source and synthesis methods are used here to classify polymers.

Condensation polymerization is polymerization in which monomers are covalently linked via the splitting off of water and/or other simple molecules. Addition polymerization is polymerization in which monomers are covalently linked without the splitting off of water and/or other simple molecules. Most addition polymerizations are based on reactions with the carbon–carbon double bonds of component monomers. Four primary methods of addition polymerization are bulk polymerization, solution polymerization, suspension polymerization, and emulsion polymerization. Table 1-3 compares addition polymerization methods.

Low Molecular Weight Polycarboxylates

Important water-soluble polycarboxylates are homopolymers of acrylic acid alone or copolymers of acrylic acid with acrylamide, methacrylic acid, or maleic acid. Linear (i.e., straight-chain) low molecular weight (i.e., less than 100,000) polycarboxylates made from acrylic acid, methacrylic acid, and/or maleic acid (Figure 1-1) are used as water-soluble anionic dispersants, flocculants, and thickening agents. For example, water-soluble polyacrylic acid dispersants of approximately 5,000 to 70,000 molecular weight have

Table 1.3. Comparison of polymerization methods

Polymerization method	Method characteristics
Bulk polymerization	As no solvent or diluent is present, the polymer is isolated as a solid that must be dissolved or dispersed before use.
Solution polymerization	The polymer is obtained as a solution that may be used as such. The method can be used to prepare vinyl, acrylic polymers, alkyds, and polyesters.
Suspension polymerization	Polymerization takes place in individual monomer droplets in a water dispersion. The polymer can be isolated by coagulation and recovered as a dry powder.
Emulsion polymerization (a) Aqueous continuous phase	Polymerization takes place in very small droplets (micelles) in a water-surfactant aqueous dispersion, sometimes yielding polymers with a higher molecular weight than in solution polymerization.
(a) Nonaqueous continuous phase	The diluent is an inert organic solvent. The polymer is hydrophobic. This method can be adapted to preparation of polymer solutions by the addition of solvents for the polymer after polymerization is complete.

replaced phosphate where phosphate in laundry products has been restricted or banned under environmental regulations.

Superabsorbent Polyacrylates

Polyacrylate superabsorbents (PAS), also known as superabsorbent polyacrylates (SAP), are part of a diverse group of high molecular weight, cross-linked polymers derived from acrylic acid. High molecular weight

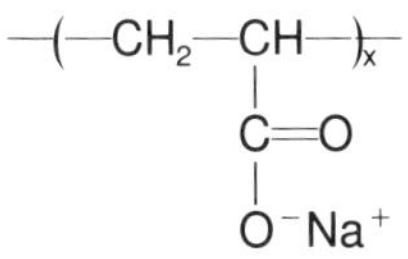

$$-(-CH_2-CH-)_x-$$
$$C=O$$
$$O^- Na^+$$

Figure 1.1. Example of a polycarboxylate functional unit.

(1 million or greater) polymers of acrylic acid are extremely sorbent with water. These polymers are found in personal hygiene and diaper products. PAS are produced by cross-linking high molecular weight polycarboxylate backbones, which carry counter ions such as sodium. The presence of carboxylate groups and the absence of any other functional groups are the characteristic features that distinguish PAS from many other absorbent materials. PAS are capable of absorbing many times their weight in aqueous fluid, depending on the ionic strength of the absorbed fluid and the properties of the PAS. For example, the fluid absorption standard for commercial-grade PAS is between 30 and 40 g urine absorbed per g dry PAS.

Water Treatment Polymers

Water treatment polymers are used by water handling facilities (e.g., wastewater or supply water treatment) for removal of solids, elimination of pollutants, and handling of sludge for solid waste disposal. Polymers that use charge neutralization to facilitate waste removal are coagulants, and polymers that function by both charge neutralization and bridging are flocculants. Both coagulants and flocculants are water-soluble high molecular weight polymers that can be broadly classified by ionic charge: cationic, anionic, nonionic, and amphoteric. Amphoteric polymers have both cationic and anionic functional groups; amphoteric charge is dependent on water pH. Structures of some commercially significant water treatment polymers are given in Figures 1-2 through 1-4.

High molecular weight cationic flocculant polymers are produced by polymerization of acrylamide and cationic monomers. These monomers can include quaternary aminoesters or aminoamides, tertiary aminoalkyl esters, and tertiary aminoalkyl amides. Some commercially important cationic flocculant products are prepared by the Mannich reaction of acrylamide with formaldehyde and dimethylamine. Other flocculants can be produced from styrene-maleic anhydride, pyridine, and polyvinylbenzyltrimethyl ammonium chloride. Anionic polymers containing acrylamide, acrylic acids, and mineral acids can have flocculant properties. Nonionic flocculants include polyacrylamides and those based on polyglycols. The molecular weight range of typical flocculants is over 1 million to approximately 50 million.

Polymeric coagulants are cationic polyamines or quaternary amines. Cationic coagulant homopolymers can be prepared by condensation reactions of ethyleneimine, vinylamine, or epichlorohydrin/tertiary amine. Other important coagulants include polymers prepared from epihalohydrins, dihaloalkanes, and aziridines, as well as hydrophobic polymers such as styrene/imidazoline copolymers. In addition, a commercially significant

Cationic Esters

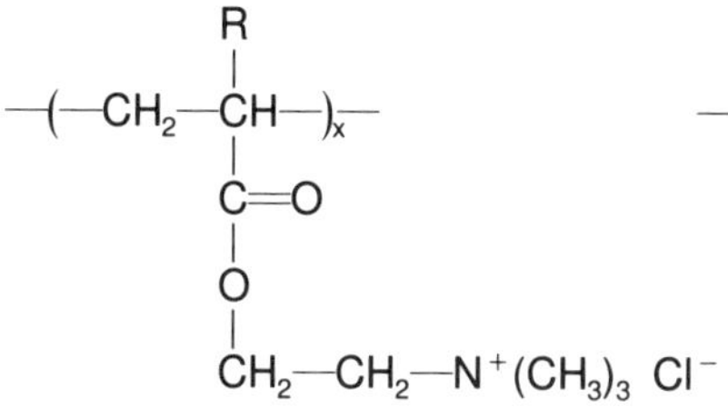

$$-(-CH_2-CH-)_x-$$

with R on the CH, and below: $C=O$, O, $CH_2-CH_2-N^+(CH_3)_3 \ Cl^-$

Cationic Amides

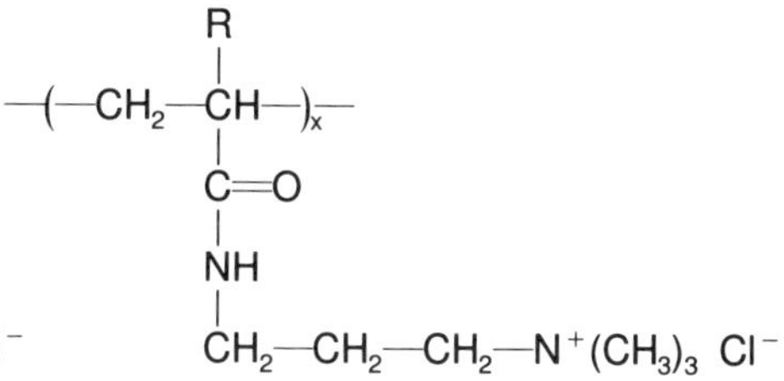

$$-(-CH_2-CH-)_x-$$

with R on the CH, and below: $C=O$, NH, $CH_2-CH_2-CH_2-N^+(CH_3)_3 \ Cl^-$

Quaternized Amines

$$-[-N^+(CH_3)_2-CH_2-CH-CH_2-N^+(CH_3)-CH_2-CH-CH_2-]-$$

with OH below first CH; $(CH_2)_2$ below the middle; OH below last CH

$NH-CH_2-CH-CH_2- \ 2Cl^-$

OH

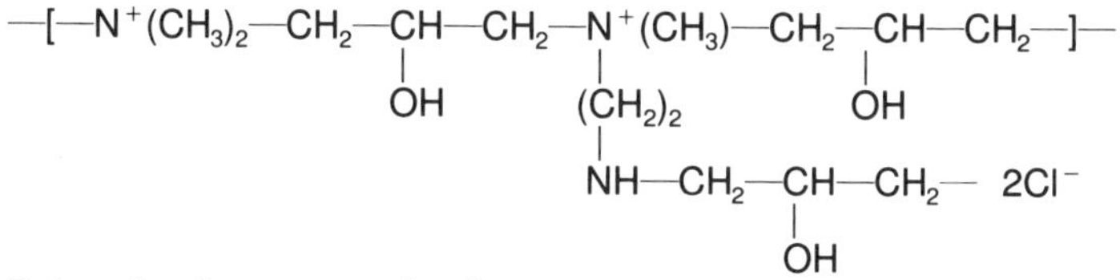

Polyamine (non-quaternized)

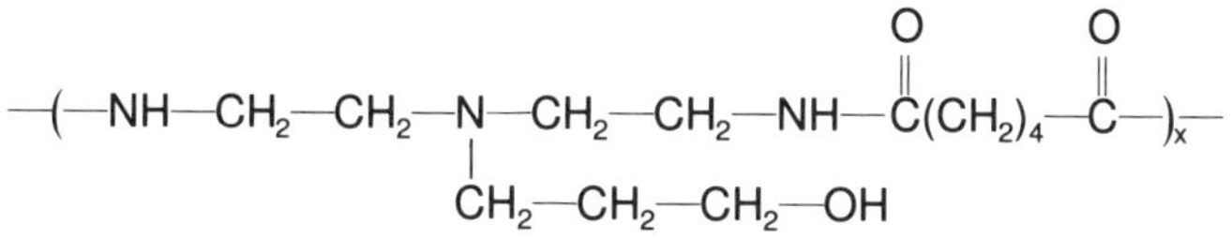

$$-(-NH-CH_2-CH_2-N-CH_2-CH_2-NH-\overset{O}{\overset{\|}{C}}(CH_2)_4-\overset{O}{\overset{\|}{C}}-)_x-$$

$CH_2-CH_2-CH_2-OH$

Cationic Pyridine

$$-(-NH-\underset{N}{C}\underset{N}{\overset{C}{\parallel}}C-NH-CH_2-NH-)_x-$$

with ring: N, C, N, C, N, C, NH

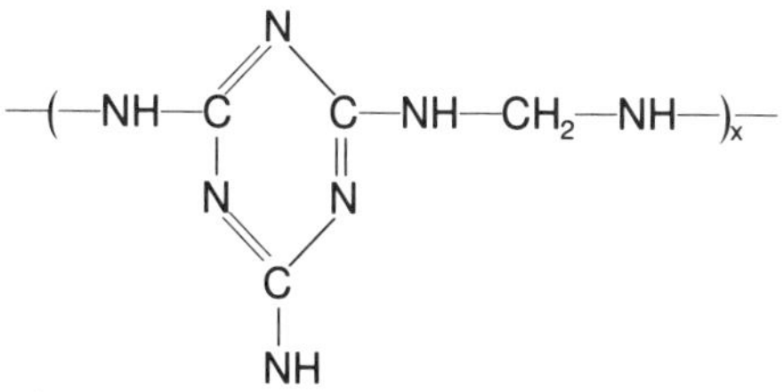

Figure 1.2. Examples of functional units of cationic water treatment polymers.

Acrylic Acid

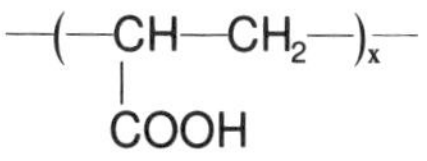

$$-(-CH-CH_2-)_x-$$

$COOH$

Acrylic Acid/Acrylamide (AA/AM)

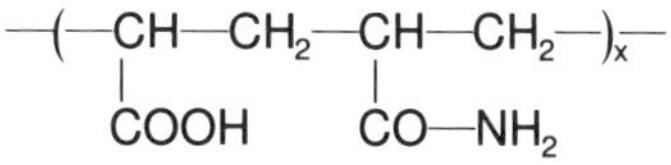

$$-(-CH-CH_2-CH-CH_2-)_x-$$

$COOH \quad CO-NH_2$

AMPS/Acrylamide (AMPS = Acrylamidomethylpropane sulfonic acid)

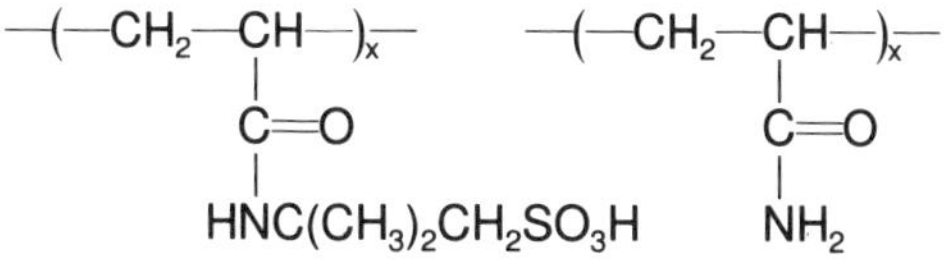

$$-(-CH_2-CH-)_x- \qquad -(-CH_2-CH-)_x-$$

$C=O$, $HNC(CH_3)_2CH_2SO_3H$ and $C=O$, NH_2

Figure 1.3. Examples of functional units of anionic water treatment polymers.

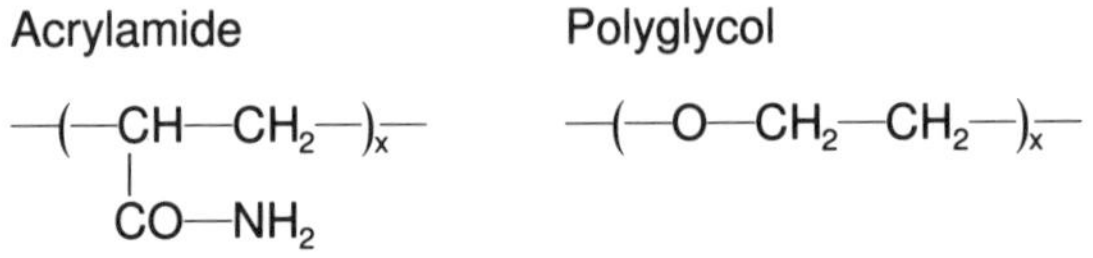

Acrylamide

$-(-CH-CH_2-)_x-$
  |
  $CO-NH_2$

Polyglycol

$-(-O-CH_2-CH_2-)_x-$

Figure 1.4. Examples of functional units of nonionic water treatment polymers.

class of coagulant polymers is based on dialkyldiallylammonium halides. Coagulants have molecular weights ranging from 10,000 to 500,000.

High Molecular Weight Suspension, Solution, and Emulsion Polymers

Many high molecular weight (i.e., greater than 100,000) polymers from suspension, solution, or emulsion polymerization are virtually insoluble in water and/or common organic solvents. These polymers are used in a wide range of finished products, including adhesives, paints, floor polishes, and oil additives. The typical molecular weight of polymers produced from suspension and emulsion polymerization processes can range from 100,000 to 3 million or more; molecular weights of many polymers from solution polymerization tend to be in the 10,000 to 100,000 range. Polyacrylic polymers (Figure 1-5) are commonly used in coating products (e.g., floor polishes and paints). Water-based dispersions of acrylic polymers can be produced by aqueous emulsion polymerization. Hydroxyl and/or carboxyl groups are introduced into acrylic polymers by use of hydroxyalkyl acrylates, acrylic acid, methacrylic acid, or itaconic acid. Other acrylic coating polymers may contain acrylamide, N-hydroxymethylacrylamide, or N-(alkoxymethyl) acrylamide. Amino, quaternary amino, and epoxy groups are also occasionally present in acrylic polymers. Styrene, vinyl toluene, and

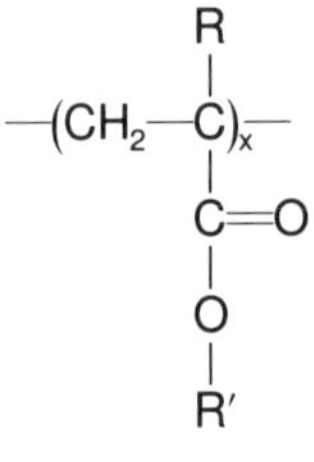

$$-(CH_2-C)_x-$$

R = hydrogen (H) or alkyl
R' = H, alkyl, hydroxy alkyl, and other functionalities (e.g., acids, amides, aminos, epoxy)

Figure 1.5. Acrylic coating polymer.

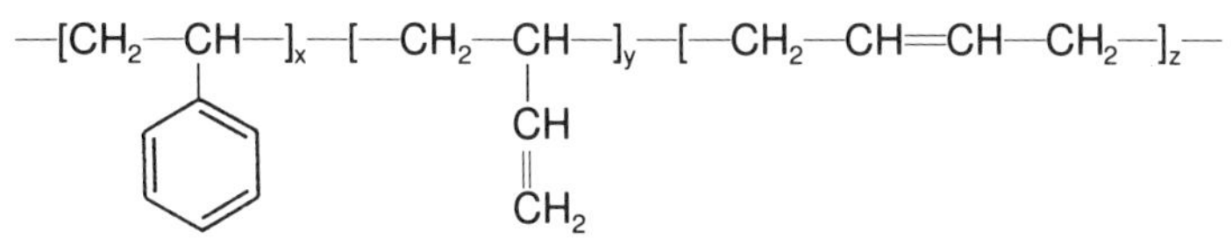

$$-[CH_2-CH-]_x-[-CH_2-CH-]_y-[-CH_2-CH=CH-CH_2-]_z-$$

Figure 1.6. Styrene–butadiene copolymer (x, y, z = number of subunits).

acrylonitrile are often included to provide specific properties. Amino-containing emulsion polymers can contain significant quantities of primary, secondary, tertiary, or quaternary nitrogen groups. The amino functional group is normally incorporated into such polymers by the use of aminoalkyl acrylate ester monomers. Amino-containing polymers are relatively reactive materials that often contain methylol or alkoxymethylol groups. Some types are based on the reaction of urea, melamine, or benzoguanamine with formaldehyde followed by reaction with a low molecular weight alcohol. Unsaturated hydrocarbon emulsion and solution polymers (Figure 1-6) are materials that contain a significant amount of ethylene-type unsaturated functionality derived from one of the hydrocarbon monomers. These are homopolymers or copolymers of diene monomers such as butadiene, isoprene or their chlorinated derivatives along with styrene, acrylonitrile, and/or other comonomers.

Products based on vinyl acetate include emulsions and solutions of polyvinyl acetate, polyvinyl alcohol, and polyvinyl acetals. Vinyl acetate copolymers (Figure 1-7) are often formed by polymerizing vinyl acetate with alkyl acrylates (commonly called vinyl acrylics), alkyl maleates and fumarates, other vinyl esters, and ethylene. Hydroxyl or carboxyl functionality is introduced via hydroxyalkyl acrylates or acrylic acid. Other vinyl acetate copolymers containing nitrogen are made by copolymerization of vinyl acetate with acrylamide or its methylol or alkoxymethyl derivatives. Additionally, amino, quaternary amino, or epoxy functionalities can be introduced. Polyvinyl alcohol is prepared by hydrolysis of polyvinyl acetate, and polyvinyl acetals are made by the acid-catalysis reaction of polyvinyl alcohol with aldehydes. Ethylene is included in some polyvinyl acetate polymers for coatings or adhesives.

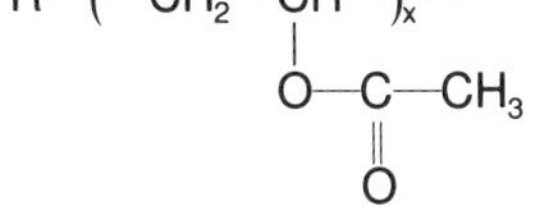

$$R-(-CH_2-CH-)_x-$$

R = monomer (e.g., acrylamide and methylolacrylamide)

Figure 1.7. Vinyl acetate copolymer.

Plastics

The most commonly known plastic polymers are the very high to infinite molecular weight polyacrylate, polyacrylonitrile, polymethacrylate, polycarbonate, polyester, polyethylene, polyisobutylene, polypropylene, polystyrene, and polyvinyl chloride products. Polycarbonates are often prepared from phosgene and bisphenol derivatives, and contain the carbonate moiety as a repeating unit. Polyester resins are normally condensation products of polybasic acids and polyhydric alcohols. Polyesters from bioengineered bacterial sources include polyalkanoates. The basic repeating unit in these polyesters is the carbonyl ester functionality. Polyalkanoates are being evaluated as biodegradable substitutes alongside natural-source cellulose and other potentially degradable plastic packaging. Both suspension and bulk polymerization have been used to produce polyethylene, polystyrene, polymethyl methacrylate, and polyvinyl chloride products.

REFERENCES

CEFIC. 1995. *Criteria for Identification of Reduced Test Package Polymers.* Brussels, Belgium: Conseil European De L'Industrie Chemique (European Chemical Industry Council).
Cheremisinoff, N. P. 1995. *Hazardous Chemicals in the Polymer Industry. New York: Marcel Dekker.*
European Union (E.U.). 1995. *Guidance Document for the Implementation of Annex VII D of Council Directive 67/548/EEC (Directive 93/105/EEC),* Document XI/584/93 rev. 2. Brussels, Belgium: European Union, Directorate-General XI.
NPCA. 1981. *Coatings Polymers: General Properties and Uses.* Washington, DC: National Paint and Coatings Association.
Odian, G. 1991. *Principles of Polymerization,* 3rd Ed. New York: John Wiley & Sons.
OECD. 1995a. *Draft Guideline: Determination of the Number-Average Molecular Weight and the Molecular Weight Distribution of Polymers Using Gel Permeation Chromatography.* Paris, France: Organisation for Economic Co-operation and Development.
OECD. 1995b. *Draft Guideline: Determination of the Low Molecular Weight Content of a Polymer Using Gel Permeation Chromatography.* Paris, France: Organisation for Economic Co-operation and Development.

OECD. 1995c. *Draft Guideline: Solution-Extraction Behavior of Polymers in Water*. Paris, France: Organisation for Economic Co-operation and Development.

Smrchek, J., R. Clements, R. Morcock, and W. Rabert. 1993. Assessing ecological hazard under TSCA: Methods and evaluation of data, in *Environmental Toxicology and Risk Assessment*. ASTM STP 1179. Philadelphia, PA: American Society for Testing and Materials.

Suter, G. W. 1993. *Ecological Risk Assessment*. Boca Raton, FL: Lewis Publishers.

U.S. EPA. 1992. *Classification Criteria for Environmental Toxicity and Fate of Industrial Chemicals*. Washington, DC: U.S. Environmental Protection Agency, Chemical Control Division, Office of Pollution Prevention and Toxics.

Chapter 2

Overview of Ecological Toxicity Test Methods

- Krzysztof M. Jop
  Science Applications International Corporation

INTRODUCTION

Ecotoxicology can be defined as the science of assessing the effects of toxic substances on ecosystems, with the goal of protecting entire ecosystems and not merely isolated components (Hoffman et al., 1995). Much attention recently has been focused on ecotoxicology in assessing ecological risks resulting from the manufacture, distribution, use, and disposal of chemicals. Data requirements in international product safety, notification, and registration laws include ecological toxicity (i.e., ecotoxicity) testing.

Although ecotoxicology includes the study of toxic effects at the cellular, individual, population, and community levels of the environment, the focus of this chapter is on testing with individual organisms. The framework for the ecological assessment of polymers can be based on a sequential (i.e., tiered) testing approach, usually beginning with tests of individual organisms (Hamilton et al., 1994; Smrchek et al., 1993).

The central concept of tiered testing is that, as testing proceeds through succeeding stages (i.e., tiers), then estimates of expected environmental concentrations and of the toxic threshold concentrations producing adverse biological effects can be made with each tier, with an increasing degree of accuracy and confidence. To make test program decisions at each tier, margins of safety are calculated by comparing the estimates of exposure to estimates of toxic threshold concentrations. Margins of safety can be used at each tier to decide whether additional testing is needed and/or to assess the degree of potential ecological risk from the intended use of the polymer. In the absence of prior information, acute toxicity laboratory tests (outlined below) with potentially sensitive organisms from relevant environmental compartments typically provide an initial "first tier" estimate of the inherent toxicity of polymers.

Toxicity tests are designed to describe a concentration–response relationship. Acute toxicity tests often focus on the measurement of lethality, whereas chronic tests focus on sublethal effects such as reproduction, growth, behavior, or biochemical effects. Measurement endpoints are values derived from toxicity tests at their termination. As noted above, an initial program for most polymers will use acute toxicity tests with aquatic and, occasionally, terrestrial organisms. Aquatic test organisms are usually fish, invertebrates, and algae. Terrestrial test organisms are typically earthworms. Relevance is determined primarily by the compartment (e.g., water, soil, sediments) where the polymer accumulates after discharge and/or disposal. For example, the hazard posed by a polymer disposed in sewage sludge is determined according to which organisms are potentially exposed to the polymer after disposal.

Results from acute toxicity tests are often used to predict maximum allowable concentrations in the environment. However, information generated from acute tests is typically limited to measurement of lethality and does not adequately predict the other potential consequences of exposure (e.g., alterations in the reproductive patterns of aquatic or terrestrial organisms during long-term exposure). Chronic toxicity tests produce data that develop more certainty in predicting the concentration of a polymer not likely to harm resident populations. Reductions in growth or reproduction success of the species tested have great ecological importance. In general, chronic toxicity tests provide an opportunity to estimate an acceptable concentration of polymer for long-term exposure.

The concentration–response relationship from the acute toxicity tests is usually characterized by the LC_{50}, the median effective concentration that is lethal to 50% of a test population. The EC_{50} — the concentration that immobilizes, inhibits growth or causes other sublethal effects in 50 percent of test organisms — is also used as an effect endpoint in tests with fish, invertebrates, and algae. The LC_{50} or the EC_{50} is typically calculated by one of four methods (i.e., binomial, moving average, probit, and Spearman-Karber). The no observed effects concentrations (NOEC) and no observed adverse effect concentrations (NOAEC) are often measurement endpoints in partial and full life-cycle tests for chronic toxicity. The NOEC is the highest concentration that has no statistically significant effect on the test organisms as compared to control organisms. In contrast, the NOAEC is the highest concentration found with no adverse effects when compared to control organisms. Determination of the minimum concentration at which an actual adverse effect is found is a key factor in assessing study results. The lowest observed effect concentration (LOEC) is the lowest concentration that has a statistically significant adverse effect. The geometric mean of the NOEC (or NOAEC) with the LOEC is described as the maximum allowable toxicant concentration (MATC).

Chronic toxicity tests are particularly useful when available information (e.g., acute toxicity) suggests that the polymer may be highly toxic and/or undergoes transformation to form toxic and/or potentially bioaccumulative intermediates. However, the number of studies conducted in polymer testing programs may be substantially reduced for polymers with adequate data available in the scientific literature or for those with suitable structure–activity correlations. Important qualitative structure–activity relationships from toxicity and fate studies with polymers are outlined in Parts 2 and 3 of this book. Also, less testing is appropriate for polymers that never reach aquatic or terrestrial environments and for those that would be present in the natural environment at low levels.

Chronic toxicity data should be reviewed together with acute toxicity values to establish the acute to chronic ratio (ACR). The ACR is valued for establishing confidence to predict "safe" concentrations. The lower the ratio, the greater the confidence that establishing limits based on acute toxicity will also protect against chronic effects. However, a high ACR does not necessarily imply an increased ecological hazard but rather may indicate that chronic effects were caused by physical-chemical attributes of the polymer at high test concentrations in the laboratory. Polymer "toxicity" may occasionally be related to a polymer dispersion problem during dosing of aquatic organisms. For example, dispersion problems may inhibit water light penetration, which is essential for growth of algae, or may clog gills at high polymer concentrations.

Relevant aquatic toxicity and exposure information is essential for defining the ecological risk of a polymer. Specifically, the ecological fate of a polymer should dictate the environmental compartment in which a polymer toxicity testing program should be focused. It is important to remember which tests should be chosen to achieve responsible ecological management goals, and that there is a tiered continuum of ecological risk assessments from crude screens to detailed quantitative estimates (e.g., Suter, 1993).

Standardized tests can provide much information that is useful for priority setting, selecting the least toxic among an array of polymers, and determining where resources should be used to reduce hazard. Standard physical (e.g., temperature, lighting conditions), media (e.g., water quality characteristics), and biological (e.g., a single life stage and feeding regime) conditions have reduced between-study variability in test results. However, standardized testing is not necessarily a suitable means to predict actual toxicity in ecosystems that may vary significantly from one area to another. In fact, although standardized tests reduce the need for test development. Additional information may be needed to predict actual ecological outcome (Cairns and Niederlehner, 1995).

OVERVIEW OF LABORATORY TESTS FOR ECOTOXICITY

Use of Laboratory Tests

Laboratory tests for ecological assessment with individual organisms are intended to provide baseline data to predict safe concentrations of polymers and for research purposes. Standardized laboratory tests can be required by regulatory agencies for product registration, labeling, shipping, or waste disposal. The use of Organization of Economic Cooperation and Development (OECD) test guidelines, where applicable, provides the opportunity for uniform interpretation and acceptance of test results on a global basis. For example, following the development of standard effects testing guidelines, a guidance document for aquatic effects assessment was recently issued by the OECD (OECD, 1995). However, to be broadly applicable, standard laboratory testing programs should follow the exact procedures published by regulatory agencies worldwide such as the European Union (E.U.), Environment Canada, U.S. Environmental Protection Agency (U.S. EPA), U.S. Food and Drug Administration (U.S. FDA), and environmental authorities in the Pacific region (e.g., Japan and Australia). Information on the specific requirements for ecological testing of polymers in the United States, Canada, the E.U., and the Pacific region is provided in Part 3. A list of ecotoxicity testing guidelines is presented in Table 2-1.

Although relatively well-defined test methodology is currently available (Adams, 1995), test guidelines may differ considerably for the same organisms. Table 2-2 presents endpoints and durations for the algal toxicity test required by a variety of organizations. The experimental conditions (i.e., duration of the test, light intensity, temperature, composition of the medium) across algal toxicity test protocols differ substantially, and will significantly alter the reported test results (Millington et al., 1988; Smith et al., 1987). The effects of these variables must be considered to generate meaningful data.

Ecotoxicity Tests

Aquatic Invertebrates and Fish

Standardized laboratory aquatic ecotoxicity tests typically utilize static, static-renewal, or flow-through systems (Adams, 1995). Polymer properties such as degradation potential help determine the choice of the test system (Jop et al., 1986). If a polymer degradation rate is less than 10% within a 4-day period, a static system is suitable. If the degradation rate is less than 10% in a 24-hr period, then a static-renewal method can be applied for testing. When degradation is greater than 10% in a 24-hr period, the flow-through system would adequately assess toxicity. These tests are usually performed with five test concentrations and a test medium control. A solvent control is also included if a solvent is used with the polymer as a carrier.

Organisms that have been used successfully in aquatic ecotoxicology include freshwater invertebrates such as planktonic crustaceans (e.g., *Daphnia pulex*, *Daphnia magna*, and *Ceriodaphnia dubia*) and freshwater fish (e.g., *Pimephales promelas* and *Oncorrynchus mykiss*). Marine organisms include invertebrates (e.g., *Mysidopsis bahia*) and fish (e.g., *Cyprinodon vatriegatus* and *Menidia spp.*).

Acute toxicity EC_{50} or LC_{50} values with fish range from measurement periods of a few hours to 14 days. The 72-hr or 96-hr acute toxicity values are most often used for regulatory purposes. The invertebrate acute value also can range from measurement periods of a few hours to several days. The 48-hr or 96-hr values are generally reported for freshwater species (e.g., daphnids and ceriodaphnids) and saltwater species (e.g., mysids).

As noted above, the fish chronic toxicity is generally reported as chronic NOEC and/or MATC values, based on sublethal effects. However, chronic toxicity is sometimes based on lethality. Chronic EC_{50} and LC_{50} values are occasionally reported. Fish chronic toxicity tests include multigenerational tests (i.e., two generations, whole life-cycle tests, and partial life-cycle tests). Fish chronic toxicity tests may last up to 28 days to 2 years. The cost and length of time required to perform full life-cycle tests have encouraged scientists to search for alternative approaches. Full life-cycle studies have been replaced for some purposes by protocols that allow life-cycle tests to be performed in much shorter periods. A common partial life-cycle test is the early life-stage (ELS) toxicity test. Partial life-cycle tests may also include the 6-day to 8-day toxicity test with *Ceriodaphnia dubia* and the 7-day to 8-day early life-stage test with *Pimephales promelas* as substitutes for longer 28-day to 62-day ELS tests. Conditions where the shortened ELS tests may not apply include testing (1) of low-solubility test substances, (2) when the test substance can be metabolized to form toxic metabolites, or (3)

Table 2.1. Regulatory guidelines and types of testing

Regulatory guideline	Type of testing
TSCA[a]	
795-120	*Hyalella azteca* flow-through acute
797-1050	*Selenastrum capricornutum* test
797-1160	*Lemma sp.* acute
797-1300	*Daphnia magna* acute
797-1310	*Gammarus sp.* acute
797-1330	*Daphnia magna* chronic
797-1400	Fish acute (freshwater and marine)
797-1520	Fish bioconcentration test
797-1600	Fish early life stage test
797-1800	Oyster shell deposition test
797-1830	Oyster bioconcentration test
797-1930	*Mysidopsis bahia* acute
797-1950	*Mysidopsis bahia* chronic
797-1970	Penaeid shrimp acute
FIFRA[b]	
Subdivision E	
Aquatic Test Guidelines	
72-1	Acute test with freshwater fish
72-2	Acute test with freshwater invertebrates
72-3	Acute test with marine organisms
72-4a	Fish early life stage
72-4b	Life cycle test with invertebrate
72-5	Life cycle test with fish
72-6	Bioaccumulation study
72-7	Simulated or actual field study
U.S. FDA[c]	
Environmental Effects	
4.01	Algal toxicity
4.08	*Daphnia magna* acute
4.09	*Daphnia magna* chronic
4.10	*Hyalella azeteca* acute
4.11	Fish acute
4.12	Earthworm subacute
CEPA[d]	
OECD 203	Fish acute test
OECD 202	*Daphnia magna* acute test
EPA 797-1050	*Selenastrum capricornutum* test
OECD 301	Ready biodegradability test
OECD[e]	
Aquatic Effects Testing	
201	*Selenastrum capricornutum* test
202-1	*Daphnia magna* acute

Table 2.1. Continued

Regulatory guideline	Type of testing
202-2	*Daphnia magna* chronic
203	Fish acute
204	Fish prolonged toxicity test
305A	Sequential static fish test
305B	Semi-static fish test
305C	Fish bioaccumulation
305D	Static fish test

Source: Modified from Adams, 1995.

[a]TSCA = U.S. Environmental Protection Agency (U.S. EPA) Toxic Substances Control Act.

[b]FIFRA = U.S. EPA Federal Insecticide Fungicide and Rodenticide Act.

[c]U.S. FDA = U.S. Food and Drug Administration.

[d]CEPA = Canadian Environmental Protection Act.

[e]OECD = Organisation for Economic Co-operation and Development.

when the test substance is a developmental/reproductive toxicant. Many polymers lack the potential for metabolic activation, membrane absorption, and developmental/reproductive effects.

As found in fish tests, invertebrate chronic toxicity tests include multi-generational tests (i.e., several generations, whole life-cycle tests, and partial life-cycle tests such as daphnid reproduction inhibition tests). The measurement periods of these tests are from 14 days to several months. Shortened ELS toxicity tests lasting 7 days with ceriodaphnids may substitute 14-day to 21-day daphnid reproduction inhibition tests. Shorter tests are best suited for water-soluble or well-dispersed polymers.

Table 2.2. Comparison of tests with *Selenastrum capricornutum*

Regulatory guideline	Duration	Endpoint
ASTM 1218	96 hours	Growth rate, biomass
CEPA 145	96 hours	Growth rate, biomass
EU L383A	72 hours	Growth rate, biomass
OECD 201	72 hours	Growth rate, biomass
TSCA 797-105	96 hours	Cell density
FIFRA 122-2	120 hours	Cell density
FDA 4.01	14 days	Maximum cell density and growth rate

Algae and Vascular Plants

The potential for a polymer to inhibit the growth of plants (e.g., phytotoxicity) is an important ecological consideration. The freshwater algae *Selenastrum capricornutum* and the marine algae *Skeletonema costatum* are commonly used in initial tests for phytotoxicity in aquatic environments. The algal toxicity value is reported as the EC_{50} value because sublethal effects on algal population growth are measured. The EC_{50} is generally reported as a 72-hr or 96-hr value. It should be noted that the algal toxicity test is a multigenerational test. Therefore, it is a chronic toxicity test, and the 96-hr NOEC should be reported with the 72-hr and/or the 96-hr EC_{50}.

Terrestrial phytotoxicity can be assessed by measuring the vegetative growth and vigor of vascular plants. Standard vascular plant toxicity tests are the seed germination/root elongation toxicity test and the early seedling growth toxicity test (Smith, 1991). Toxicity testing with vascular plants can also involve screening mature plants for sensitivity to test substances. Seed germination and early seedling growth tests are most often conducted with lettuce, cabbage, wheat, oats, and tomatoes. Sublethal toxic effects are generally reported as EC_{50} values. Test measurement periods of early seedling growth tests depend upon species characteristics (e.g., time to germination and growth rate). If a test is extended over a test plant's life-cycle, chronic values for that plant may be obtained. For example, entire vascular plant macrophyte toxicity tests may last up to 30 days. If the test exposure period is hours to several days, then the results should be regarded as representative of acute toxicity. However, if the exposure period is a significant portion of the life-cycle of the macrophyte, then the results of the test may be reported as chronic endpoints.

Sediment Organisms

Sediment (e.g., benthic) invertebrates include crustaceans, oligochaetes, and insects. The U.S. EPA has released a guide for conducting sediment tests with invertebrates (U.S. EPA, 1994). Commonly used organisms for freshwater sediment testing are *Chironomus tentans*, *Chironomus riparius*, and *Hyalella azteca*. *Rhepoxynius abronius* and *Ampelisca abdita* are used for marine sediment testing. As with water-column studies, toxicity with these organisms can be evaluated in acute, partial-life cycle, or full life-cycle exposure. It should be noted that toxicity tests with benthic organisms may be done with and without sediments. When sediments are included, toxicity values and exposure concentrations are typically milligrams per kilogram (i.e., milligrams of test substance per kilogram of dry weight sediment).

The toxicity test with bivalves such as oysters generally involves measurement of shell deposition or lethality to oyster larvae. Although reduction of shell deposition is a sublethal effect reported as a EC_{50}, measurements with larvae may include lethality. Test duration for the shell deposition test is typically 96 hr, whereas test duration with larvae may be up to several days.

Sediment toxicity tests can be used to further assess ecosystem stress (Burton and Scott, 1992). Physical-chemical properties such as solubility and adsorptivity, as well as ecological factors such as test sediment composition, particle size and organic carbon content, can affect a polymer's partitioning behavior (Landrum et al., 1985). Benthic organisms may be exposed to polymers that are either sorbed to sediment or in the sediment pore water. However, adsorption to particulates and complexation with other organic compounds can strongly mitigate (i.e., reduce) polymer toxicity to aquatic organisms (Lewis and Wee, 1983). Therefore, sediment toxicity testing of polymers may be relevant only in cases where bioavailability to sediment organisms is plausible.

Terrestrial Invertebrates

Earthworms are commonly used as organisms of choice in initial ecological assessments of polymer toxicity to invertebrates in soils. Over 20 species of earthworms can be found in North American soils (Olson, 1928). *Eosinia foetida* is a typical species of earthworm that is used in standard test procedures (Roberts and Dorough, 1985). Tests are generally done with natural or artificial soil, and may last from several days to 14 days. Lethality is the primary measured effect, but sublethal effects can be measured. Toxicity values and exposure concentrations are generally reported as milligrams of test substance per kilogram of dry weight soil.

Microorganisms

Toxicity to bacteria is reported with either LC_{50} or EC_{50} values (depending on the endpoint measured) and NOEC values. EC_{50} values are used when sublethal effects on population metabolism are measured (e.g., inhibition of oxygen consumption or population growth). Tests with natural soil communities and natural soils are preferred when compared to single species cultures in artificial growth media. Most tests are sufficiently long in duration (i.e., days) to cover many generations; therefore, these tests can be considered to predict chronic toxicity. More information on testing with soil microorganisms can be found in Chapter 3.

Birds and Mammals

For the purpose of ecological assessment programs with polymers (see Part 2), toxicologists have generally not included testing with birds and mammals. However, tests with mammals have been used to prepare toxicological profiles for human health assessment. The low oral toxicity profile found with polymers in mammals has probably limited interest in testing birds.

Acute oral toxicity testing with birds and mammals generally involves a single oral dose followed by 14 days of observation. Death is the primary endpoint measurement, and LD_{50} values are based on body weight (i.e., milligrams of test substance per kilogram of body weight). Acute dietary toxicity testing with birds has involved exposure to food containing test substances for 5 days followed by 3 or more days of observations for toxic effects and lethality. LC_{50} values are often based on a concentration of the substance in food (i.e., milligrams of test substance per kilogram of dry weight food). Chronic toxicity testing of birds and mammals generally involves measuring developmental effects, reproductive potential (both birds and mammals), or carcinogenesis (mammals). Chronic studies generally involve continuous exposure to contaminated food or dosing by gavage. Toxicity values are reported either by food consumption (i.e., milligrams of test substance per kilogram of dry weight food) or by body weight (i.e., milligrams of test substance per kilogram of body weight).

BIOCONCENTRATION

Bioconcentration is an initial measure of the potential for accumulation of chemical residues in a food chain (McCarty and Mackay, 1993). The bioconcentration factor (BCF) is used to quantify the magnitude of bioconcentration. In aquatic bioconcentration studies, the BCF is the proportionality between the tissue concentration of a chemical in an organism and the chemical's concentration in water, under equilibrium conditions. Procedures used to estimate BCF values are the equilibrium method, kinetic modeling methods, and quantitative structure–activity relationship (QSAR) methods. QSAR methods estimate the BCF from physical-chemical properties such as solubility in water or octanol. Factors that strongly influence bioavailability include exposure concentration, pH, the presence of dissolved organic matter, and steric hindrance. Steric hindrance has prevented membrane absorption of chemicals with large molecular size or shape (e.g., Vieth et al., 1979). Virtually all synthetic commercial polymers are 1,000 to several million in molecular weight. Therefore, BCF measurements are generally understood not to be relevant for ecological assessments of synthetic commercial polymers.

METHODOLOGICAL ISSUES

Dosing and Equilibration

Most standard test guidelines recommend that test substance concentrations be measured at the beginning and the end of the exposure. The test substances are usually dissolved in water or solvent in relatively high concentrations (i.e., stock solutions) and added directly to the test chambers. The chambers are mixed to equilibration for minutes to hours. After equilibration, testing is initiated. The equilibration time may have a profound impact on the results of the toxicity test. Before initiation of the test, several analytical measurements should be performed to assure equilibration of the test concentration.

Methods for ecotoxicity testing generally have been designed for chemicals tested at concentrations below their limit of solubility in water. However, some polymers (e.g., lubricant polymers in oil and emulsion polymers in aqueous suspension) are extremely insoluble. These and other polymers cannot effectively be extracted from the external carrier phase. Some polymer mixtures may form partial emulsions in aqueous media even in the absence of suitable dispersant conditions. To simulate actual conditions of exposure in water, water-soluble fractions or water-accommodated fractions have been considered for complex mixtures. Water-soluble fractions are prepared by complete elimination of dispersant conditions by dilution, chemical extraction, and/or filtration, followed by equilibration in water. These stock fractions may be serially diluted prior to dosing for aquatic toxicity tests. Water-accommodated fractions are prepared by equilibration in water, with filtration and chemical extraction virtually avoided. Exposures to water-accommodated fractions may include undissolved but dispersed components as well as dissolved components. Water-accommodated fractions are prepared by adding ratios of the whole polymer mixture to water. The ratios are not serially diluted, in order to avoid dilution-dependent differences in component equilibrium potential. Instead, equilibrium is established after addition of the ratios to the appropriate volumes of the test medium. The water-accommodated fraction is drawn off after a period of settling and then used as the exposure medium for testing.

It should be noted that solvents may be used, but only to assist in the addition of the mixture to the test medium, not to aid in dissolution of the mixture.

By avoiding chemical extraction and filtration, water-accommodated fractions may be a means to dose mixtures with limited solubility (Girling et al., 1994). However, the intensive resources required for preparation and analytical validation of water-soluble or water-accommodated fractions are probably not justified if the equilibrium concentration of the test polymer is below detection in terms of chemical analysis and effects testing. Under regulatory programs, justification for the testing of polymers with limited solubility by using polymer extracts, water-soluble fractions, or water-accommodated fractions should be based on a balance between reasonable cost and expected levels for polymer toxicity.

Phytotoxicity Measurement Alternatives

The use of counting techniques to determine toxicity to algae is relatively simple, but extremely time-consuming. The use of more rapid, less labor-intensive technologies may be more appropriate than counting. Growth tests for plants have been adopted to reduce labor costs, and a rapid algal toxicity test is based on measurement of photosynthesis. This can be accomplished by monitoring photosynthesis with carbon dioxide uptake, oxygen evaluation, or fluorescence emission characteristics (Smith, 1991).

Sediment Sampling

Two types of sediments can be used in sediment toxicity studies: a field-collected sediment and a formulated "reference" sediment. It might appear that a reference sediment would be best for testing polymers. Formulated reference sediments were developed to simulate the physical-chemical characteristics of field-collected sediments (Suedel and Rodgers, 1994). However, natural field sediments are complex mixtures comprised of minerals, organic materials, water, and a diverse population of microorganisms (Burton and MacPherson, 1995). Bacteria fauna profiles change with depth in the sediment core sample. Therefore, reference sediments are used with caution in standard testing; field-collected sediments are most commonly used for testing chemicals.

Little is known about the changes that occur in a sediment after sample collection and processing but before toxicity testing. Sediments are often removed from the core sample container for sieving and mixing to remove indigenous organisms; but with sediment homogenization, natural redox stratification and microbiological gradients are disrupted. Further, sufficient time should be allowed before initiation of the toxicity test for equilibration of the introduced polymer with the sediment.

Sediment test samples must be virtually free of contaminants. Background information is essential to characterize the sediment. Relevant characterization parameters include total organic carbon content, particle size distribution, pH, ammonia, and contaminant levels (e.g., metals and chlorinated and petroleum hydrocarbons).

The method of introducing a polymer to a sediment depends on the polymer's physical-chemical characteristics. Water-soluble or readily emulsifiable polymers can be easily prepared in a small aliquot of water and then mixed with the sediment. With poorly water-soluble polymers, a common approach is wet mixing of the polymer and the sediment. These polymers are dissolved first in an organic solvent such as acetone and then mixed with the sediment. Before further handling, the polymer–sediment mixture may be added to a container to allow the carrier to evaporate. Prior to addition of the overlying water column and sediment test organisms, the polymer–sediment mixture is left for 24 hr to 14 days to establish equilibrium.

Testing of the interstitial pore water of sediments may produce the most consistent and representative results. Measurement of sediment toxicity with interstitial pore water rather than whole sediment is based on the assumption that organisms receive most of their exposure through contact with the sediment's interstitial water (Knezovich et al., 1987). There are a number of methods to extract interstitial pore water from the sediment, including squeezing under pressure and centrifugation. All the laboratory methods change the interstitial water chemistry, but centrifugation may alter it the least. Toxicity testing using pore water can utilize sensitive assays such as the short-term chronic test with *Ceriodaphnia dubia*.

There has been concern over the use of surrogate organisms such as *C. dubia* as indicators of indigenous species effects at the same or differing levels of biological organization. Information presented by the U.S. EPA (1987) has suggested that sensitive laboratory organisms and test methods will show responses that are sufficiently protective for adverse effects in resident fish and macroinvertebrate populations.

Polymer concentration in sediment interstitial pore water will not always reflect the total bioavailable fraction of the polymer in the sediment. Some polymers that are tightly bound to sediment particulates will affect only organisms that actively feed using these particulates. The toxicity of the polymers that reach equilibrium in the interstitial pore water may be better evaluated by using benthic invertebrates with particulate feed behavior, such as midge larvae or oligochaetes.

QUALITY ASSURANCE AND CONTROL

Polymer toxicity testing programs, particularly those used for regulatory purposes, follow Good Laboratory Practice (GLP) regulations. These regulations are largely based on FIFRA (Garner et al., 1992) and OECD GLP principles and guidelines (OECD, 1992a–e, 1993a, b, 1994).

GLP regulations for laboratory practice are typically divided into sections that describe requirements for the facility and its personnel, study conduct, the responsibilities of the study director and the quality assurance unit, requirements for protocols, standard operating procedures (SOP), reporting, and archiving. GLPs require that all staff members have adequate education, training, or experience to perform their assigned tasks. The testing facility should adequately separate culture, administrative, and testing areas to prevent cross-contamination. SOPs must be followed, and, for each test, a specific protocol should be written that describes the study plan, methods to be employed, and data to be recorded and reported.

Quality assurance programs include quality control procedures. Quality control includes the specific actions for obtaining prescribed standards of performance as part of a quality assurance program. These actions include standardization, calibration, replication, and control and reference samples suitable for providing statistical estimates of data confidence. GLP regulations can also apply to measurement of the fate of polymers in the environment. These studies are detailed in Chapter 3.

REFERENCES

Adams, W. J. 1995. Aquatic toxicology testing methods, in *Handbook of Ecotoxicology*, ed. D. J. Hoffman, B. A. Rattner, G. A. Burton, Jr., and J. Cairns, Jr. Boca Raton, FL: Lewis Publishers.

Burton, G.A., Jr. and C. MacPherson. 1995. Sediment toxicity testing issues and methods, in *Handbook of Ecotoxicology*, ed. D. J. Hoffman, B. A. Rattner, G. A. Burton, Jr., and J. Cairns, Jr. Boca Raton, FL: Lewis Publishers.

Burton, G. A. and K. J. Scott. 1992. Sediment toxicity evaluations: Their niche in ecological assessments. *Environmental Science and Technology* 26:2068–77.

Cairns, J. Jr., and B. R. Niederlehner. 1995. *Ecological Toxicity Testing*. Boca Raton, FL: Lewis Publishers.

Girling, W. L., M. S. Barge, J. P. Ussary. 1992. *Good Laboratory Practice Standards.* Washington, DC: ACS Professional Reference Book.

Girling, A. E., G. F. Whale, and D. M. M. Adema. 1994. A guideline supplement for determining the aquatic toxicity of poorly water-soluble complex mixtures using water-accommodated fractions. *Chemosphere* 29:2645–49.

Hamilton, J. D., K. H. Reinert, and M. B. Freeman. 1994. Aquatic risk assessment of polymers. *Environmental Science and Technology* 28:186A–92A.

Hoffman, D. J., B. A. Rattner, G. A. Burton, Jr., and J. Cairns, Jr. 1995. Introduction, in *Handbook of Ecotoxicology,* ed. D. J. Hoffman, B. A. Rattner, G. A. Burton, Jr., and J. Cairns, Jr. Boca Raton, FL: Lewis Publishers.

Jop, K. M., J. H. Rodgers, Jr., E. E. Price, and K. L. Dickson. 1986. Renewal device for test solutions in *Daphnia* toxicity tests. *Bulletin of Environmental Contamination and Toxicology* 36:95–106.

Knezovich, J. P., F. L. Harrison, and R. G. Wilhelm. 1987. The bioavailability of sediment-sorbed organic chemicals: A review. *Water, Air and Soil Pollution* 32:233–45.

Landrum, P. F., M. D. Reinhold, S. R. Nihart, and B. J. Eadie. 1985. Predicting the bioavailability of organic xenobiotics to *Pontoporeia hoyi* in the presence of humic and fulvic materials and natural dissolved organic matter. *Environmental Toxicology and Chemistry* 4:459–68.

Lewis, M. A. and V. T. Wee. 1983 Aquatic safety assessment for cationic surfactants. *Environmental Toxicology and Chemistry* 2:123–31.

McCarty, L. S. and D. Mackay. 1993. Enhancing ecotoxicological modeling and assessment. *Environmental Science and Technology* 27:1719–29.

Millington, L. A., K. H. Goulding, and N. Adams. 1988. The influence of growth medium composition on the toxicity of chemicals to algae. *Water Research* 22:1593–602.

OECD. 1992a. *Principles of Good Laboratory Practice.* OECD Environment Monograph No. 45. Paris, France: Organisation for Economic Co-operation and Development.

OECD. 1992b. *Guides for Compliance Monitoring Procedures for Good Laboratory Practice.* OECD Monograph No. 46. Paris, France: Organisation for Economic Co-operation and Development.

OECD. 1992c. *Guidance for the Conduct of Laboratory Inspections and Study Audits.* Environment Monograph No. 47. Paris, France: Organisation for Economic Co-operation and Development.

OECD. 1992d. *Quality Assurance and GLP.* Environmental Monograph No. 48, Paris, France: Organisation for Economic Co-operation and Development.

OECD. 1992e. *Compliance of Laboratory Suppliers with GLP Principles.* Environment Monograph No. 49. Paris, France: Organisation for Economic Co-operation and Development.

OECD. 1993a. *The Application of the GLP Principles to Short-Term Studies.* Environment Monograph No. 73. Paris, France: Organisation for Economic Co-operation and Development.

OECD. 1993b. *The Role and Responsibilities of the Study Director in GLP Studies.* Environment Monograph No. 74. Paris, France, Organisation for Economic Co-operation and Development.

OECD. 1994. *The Application of the GLP Principles to Field Studies.* Environment Monograph No. 50. Paris, France: Organisation for Economic Co-operation and Development.

OECD. 1995. *Guidance Document for Aquatic Effects Assessment.* OECD Environment Monograph No. 92. Paris, France: Organisation for Economic Co-operation and Development.

Olson, H. W. 1928. The earthworms of Ohio. *Ohio Biological Survey* 4:46–55.

Roberts, B. L. and H. W. Dorough. 1985. Hazards of chemicals to earthworms. *Environmental Toxicology and Chemistry* 4:307–16.

Smith, P. D., D. L. Brockway, and F. E. Stancil. 1987. Effects of hardness, alkalinity and pH on the toxicity of pentachlorophenol to *Selenastrum capricornutum* (Printz). *Environmental Toxicology and Chemistry* 6:891–901.

Smith, B. M. 1991. An inter- and intra-agency survey of the use of plants for toxicity assessment, in *Plants for Toxicity Assessment*, 2nd Volume, ASTM STP 1115. Philadelphia, PA: American Society for Testing and Materials.

Smrchek, J., R. Clements, R. Morcock, and W. Rabert. 1993. Assessing ecological hazard under TSCA: Methods and evaluation of data, in *Environmental Toxicology and Risk Assessment.* ASTM STP 1179. Philadelphia, PA: American Society for Testing and Materials.

Suedel, B. C. and J. H. Rodgers, Jr. 1994. Development of formulated reference sediments for freshwater and estuarine sediment testing. *Environmental Toxicology and Chemistry* 13:1163-72.

Suter, G. W. III, 1993. *Ecological Risk Assessment.* Chelsea, MI: Lewis Publishers.

U.S. EPA. 1987. *Biomonitoring to Achieve Control of Toxic Effluents.* Washington, DC: U.S. Environmental Protection Agency.

U.S. EPA. 1994. *Methods of Measuring the Toxicity and Bioaccumulation of Sediment-Associated Contaminants with Freshwater Invertebrates.* Washington, DC: U.S. Environmental Protection Agency.

Vieth, G. D., D. L. DeFoe, and B. V. Bergstedt. 1979. Measuring and estimating the bioconcentration factor of chemicals in fish. *Journal of the Fish Research Board of Canada* 36:1040–1048.

Chapter 3

Overview of Ecological Fate Test Methods

■ Jon Doi
Roy F. Weston Inc.

Introduction

The potential for transport and accumulation of a substance in the environment can be estimated from measurements obtained in ecological fate studies. Measurement of ecological fate helps to predict how a polymer will be distributed and at what concentration in the natural environment. Properties that can help predict the ecological fate of a polymer include water solubility, molecular weight distribution, acid dissociation constant(s), degradation half-life values, and organic adsorption–desorption data.

Some polymers have the potential to be transformed in the environment by both biodegradation and chemical degradation mechanisms (e.g., hydrolysis and photolysis). Biodegradation can occur under aerobic (i.e., with oxygen) and anaerobic (i.e., without oxygen) conditions. Aerobic biodegradation testing is divided into three categories: ready, inherent, and simulation. Ready biodegradability tests are less favorable to biodegradation in that low concentrations of biodegrading inoculum is placed in the test system, and the only source of carbon for biodegradation is the polymer. Inherent biodegradability tests are more favorable than ready biodegradability tests for biodegradation. Higher levels of inoculum that may be acclimated to the polymer are used, and glucose can be added to the test system so that cometabolism may take place. Simulation tests are laboratory tests intended to represent actual unit operations of wastewater treatment plants (WWTP). All anaerobic biodegradation tests are considered inherent tests.

There are several reasons for conducting ecological fate tests with polymers:

- To satisfy regulatory compliance issues under the U.S. Environmental Protection Agency Toxic Substances Control Act (U.S. EPA TSCA), when submitting a New Drug Application (NDA) under U.S. Food and Drug Administration (U.S. FDA) guidelines, or when following regulations for the European Union (E.U.) and the Canadian Environmental Protection Act (CEPA).
- To submit an ecological risk assessment or environmental impact statement to a federal, state, or local regulatory agency
- For ecolabeling (i.e., environmental labeling) issues. The E.U. has taken the lead in ecolabeling. The potential to avoid an adverse product label is a powerful incentive to perform ecological fate testing. In the United States, ecolabeling currently takes on a somewhat different perspective. Nongovernmental trade organizations may evaluate the biodegradability of a chemical and, if it is found to be biodegradable, a commercial ecolabel may be placed on the product. Other countries are beginning to adopt regulatory-test-based programs for ecolabeling, including Canada and Germany.
- For marketing purposes (i.e., to differentiate one's own product from another based on biodegradability).
- For assessment of use.

This chapter provides an overview of ecological fate methods that have been used with polymers, focusing on important methods used to assess the fate of polymers in wastewater. There are other important methods for ecological fate testing of polymers for environmental degradability, which are discussed in detail in Chapters 4 and 5 on the biodegradation testing of plastics in soil and compost.

OVERVIEW OF REGULATORY TEST GUIDELINES

Aquatic and Wastewater

Ecological fate testing has focused primarily on fate in aquatic environments and/or wastewater treatment systems. The aqueous environment is a common route for disposal of economically important water-soluble consumer and industrial polymers. Aerobic wastewater microorganisms are the most common inoculum for biodegradation fate testing. Whereas there is a single guideline for an anaerobic biodegradability test and one soil biodegradation test, there are no fewer than ten regulatory guideline aqueous aerobic biodegradation tests. The major regulatory test guidelines that are followed in ecological fate test programs are those from OECD (1993a), U.S. EPA (1994), U.S. FDA (1987), and the European Union (1992).

Descriptions of relevant physical-chemical properties of polymers and their ecological significance are given in Table 3.1. These physical-chemical properties help define how a polymer will be transported, will accumulate, and will chemically degrade in the environment. Physical and chemical properties can be used in helping to assess ecological risk. A comparison of various regulatory methods for physical-chemical properties is shown in Table 3.2. Typical analytical measurements for each method are also given in Table 3.2. A comparison of biodegradation methods and associated analytical measurements, plus the criteria for "passing" the various tests, is provided in Table 3.3.

Properties that are specific to polymers are better suited to predict the ecological fate of polymers than some traditional measurements used for monomeric substances. Individual polymers are often a distribution of polymer chains with different molecular weights, not a single molecular weight. Molecular weight distribution can influence the overall degree of biodegradability and partitioning characteristics of the total polymer. Water solubility and biodegradation potential often decrease with increasing polymer chain length. Further, synthetic polymers tend to be of low volatility in air and will generally not be absorbed through biological membranes to bioaccumulate in tissues. Therefore, most regulatory authorities either do not require measurement of octanol–water partition coefficients (K_{ow} or P_{ow}) and vapor pressures of polymers (e.g., United States, Australia). Under specific circumstances, K_{ow} measurements can be required with new polymers in Canada, but these measurements may be waived (see Chapter 11). Water solubility, molecular weight, dissociation constants, sorption and desorption, hydrolysis, photodegradation, and charge density are more important predictors of the ecological fate and ecotoxicity of polymers than K_{ow} and vapor pressure.

Within test categories, standardized fate test methods are very similar. However, one important difference is that U.S. FDA biodegradation methods require three replicates, whereas other regulatory agencies may require only one or two replicates. Another difference is that optional toxicity and abiotic controls can be added to the test design in the 1994 version of the OECD/E.U. carbon dioxide (CO_2) ready biodegradability

Table 3.1.   Ecological importance of physical-chemical properties of polymers

Property	Property description	Ecological importance
Water solubility	The maximum amount of a polymer in solution and at equilibrium with excess compound in water at specified environmental conditions (i.e., temperature, atmospheric pressure, and pH).	This property, with others, governs the tendency of a polymer to move and be distributed between the various environmental compartments. Water-soluble polymers are more likely to be transported and distributed by the hydrologic cycle than are water-insoluble polymers.
Molecular weight	Number-average molecular weight (M_n) is the total weight of all the molecules in a polymer sample divided by the total number of moles present. Weight–average molecular weight (M_w) is the mean of the weight distribution of molecular weights. Polydispersity (Pd) is defined as the breadth of the distribution of molecular weights in a polymer (M_w/M_n).	Polymers are often not a single molecular weight, but a distribution of molecular weights. M_n, M_w, and Pd influence the degree of solubility, biological absorption, degradability, and adsorption potential of the entire polymer molecular weight distribution.
Dissociation constant(s)	Measure of the degree of ionization of a polymer, which varies with the pH of the solution.	The distribution of a polymer in the environment (i.e., ionized or nonionized) is partly a function of the pK of the polymer and the pH of the environment in which the polymer is found. This will affect the availability of the polymer to enter into physical, chemical, and biological reactions.
Sorption and desorption	Sorption is the adhesion of molecules to surfaces of solid bodies with which they are in contact. Partitioning refers to the dissolving of molecules into the (amorphous) solid phase. Partitioning or adsorption (or both) may account for the sorption of neutral and charged organic molecules to soil and sediment. Desorption is the reverse process of sorption.	These properties help evaluate the transport tendencies of polymers at the soil–water and soil–air interfaces. The sorption and the desorption of polymers in soil and sediment will influence their potential for leaching into the water table, moving into runoff water, or evaporating from soil into air. It will also influence a polymer's potential to photodegrade.

Table 3.1. Continued

Property	Property description	Ecological importance
Hydrolysis	Reaction of polymer (RX) with water (HOH), with the resultant net exchange of a group (X) from the polymer for the OH group from water at the reaction center as shown: $RX + HOH \leftrightarrow ROH + HX$.	Hydrolysis is one of the most common reactions occurring in the environment. Though hydrolysis rates depend on pH, temperature, and polymer concentration, they are independent of many other factors that normally affect other degradative processes (i.e., amount of sunlight, microbial population, and oxygen supply).
Photodegradation	Process whereby polymers are altered directly as a result of irradiation (by absorption of ultraviolet or visible light, which can transform the molecule into one or more products) or indirectly through interaction with products of direct irradiation (as either a catalyst or a reactive species).	Atmospheric photodegradation involves primarily indirect mechanisms, principally the interaction of a polymer with the reactive species, hydroxyl radicals, and ozone. In many instances, photo-degradation provides the same products as metabolism by plants and microorganisms, but can also "open up" a recalitrant polymer and cause acceler-ated disappearance of the polymer from the environment.
Charge density	Proportional weight of cationic (e.g., quaternary ammonium) or anionic e.g., carboxylate) fragments in the polymer chain.	Polymers containing charged functionalities may cause toxicity to aquatic organisms, possibly by interacting with exposed biological membranes such as gills and cells walls and/or by sequestering essential nutrients.
Vapor pressure	The force per unit area exerted by a gas in equilibrium with its liquid or solid phase at a specific temperature. It can be thought of as the solubility of a substance in air and is dependent on the nature of the compound and temperature.	This property governs the tendency of a vapor to be transported in air. However, vapor pressures of polymers tend to be negligible.

Table 3.1. Continued

Property	Property description	Ecological importance
Octanol/water partition coefficient	Ratio of the concentrations of any single molecular species in two phases, *n*-octanol and water, when the phases are in equilibrium with one another and the substance is in dilute solution in both phases.	This property indicates the tendency of a nonionized organic chemical to accumulate in lipid tissue and to sorb onto soil particles or onto the surface of organisms or other particulate matter coated with organic material. Polymers will not tend to be absorbed through biological membranes; therefore, octanol/water measurements with polymers may not be ecologically relevant.

test, whereas the U.S. versions do not have such controls. However, as noted above, all biodegradation test guidelines can be separated into three general test categories: ready, inherent, and simulation. A description of each is given below.

Ready Biodegradability

Ready biodegradability is a term used to indicate that a polymer can undergo rapid and ultimate biodegradation in an aerobic aquatic environment under very stringent test conditions (i.e., low concentration of biomass, no acclimation of the biomass to the polymer, and the only source of carbon as food for the inoculum being the polymer). Mineralization is the breakdown of an organic chemical to CO_2, water, and oxides or mineral salts of other elements. The endpoint for a ready biodegradability test can be CO_2 production, dissolved organic carbon (DOC), biochemical oxygen demand (BOD), or dissolved oxygen, as shown in Table 3.3. The measurement of CO_2 production is considered the most representative demonstration of ready biodegradability because there is often a direct correlation between CO_2 produced and ultimate biodegradation. All other measurements only imply ultimate biodegradation.

It is important to note that a low ready biodegradability result does not mean that the polymer has no potential for biodegradation. The polymer's potential for biodegradation under less stringent test conditions can be explored by performing other biodegradability tests such as inherent, anaerobic, simulation, or soil tests.

Table 3.2. Physical-chemical tests by regulatory agencies[a]

| | Test Method Reference Numbers | | | | |
Test type	U.S. EPA TSCA	U.S. FDA	OECD	E.U.	Analytical measurement
Water solubility	796.1840 or 796.1860	3.01	105	A.6	^{14}C or chemical-specific method
Molecular weight	n.a.[b]	n.a	Draft[c]	n.a	Chemical-specific
Dissociation constant(s) in water	796.1370	3.04	112	n.a.	Chemical-specific
Sorption and desorption	796.2750	3.08	106	n.a.	^{14}C or chemical-specific
Hydrolysis as a function of pH	796.3500	3.09	111	C.7	Chemical-specific
Photodegradation	796.3700	3.10	n.a.	n.a.	Chemical-specific
Octanol/Water[d] partition coefficient	796.1550, 796.1570, or 796.1720	3.02	107	A.8	^{14}C or specific chemical
Vapor pressure[d]	796.1950	3.03	104	A.4	^{14}C or specific chemical

[a]Methods from current guidelines may not apply to all polymers. For example, OECD is drafting a guideline on solution-extraction of polymers in water (OECD, 1995c).

[b]n.a. = not available.

[c]See OECD 1995a, b.

[d]See text and Table 3.1 for discussion of ecological relevance.

Aerobic Inherent Biodegradability

Inherent biodegradability means that a polymer can be degraded under conditions that are highly favorable to the selection and/or the adaptation of microorganisms capable of biodegradation. These methods involve exposure of the polymer to relatively high microorganism concentrations over a relatively long period of time. The viability of the microorganisms is maintained over this period by daily addition of a settled sewage feed. Because of the long sludge retention times and the daily addition of nutrients to the test system, these tests do not simulate conditions found at a wastewater treatment plant. However, a positive result in these tests

Table 3.3. Biodegradation tests from regulatory authorities

Test type	U.S. EPA TSCA	U.S. FDA	OECD	E.U.	Test criteria[a]	Analytical measurement
Ready Tests						
1. Closed Bottle	796.3200	not available	301D	C.4-E	>60% of ThOD[b]	Dissolved oxygen
2. Modified MITI	796.3220	not available	301C	C.4-F	>60% BOD, 70% LPC[c]	Oxygen consumption
3. Modified OECD Screening	796.3240	not available	301E	C.4-B	>70% loss of DOC	Dissolved organic carbon
4. Modified Sturm	796.3260	3.11	301B	C.4-C	>60% of theor. CO_2	CO_2 production
5. DOC Die-Away	not available	not available	301A	C.4-A	>70% loss of DOC	Dissolved organic carbon
6. Manometric Respiratory	not available	not available	301F	C.4-D	>60% of ThOD	Oxygen consumption
Inherent Tests						
1. Modified SCAS	796.3340	not available	302A	not available	>20%/70% DOC loss[d]	Dissolved organic carbon
2. Modified Zahn Wellens	796.3360	not available	302B	not available	>20%/70% DOC loss	Dissolved organic carbon

Simulation Tests

1. Coupled Units	796.3300	not available	303A	not available	Dissolved organic carbon
2. Porous Pot	considered[e]	not available	considered[e]	considered[e]	Dissolved organic carbon
3. CAS	not available	not available	not available	not available	Dissolved organic carbon

Miscellaneous Tests

Biodegradability in Seawater	not available	not available	306	not available	Dissolved organic carbon or dissolved oxygen
Inherent Biodegradability in Soil	796.3400	3.12	304A	not available	$^{14}CO_2$ production
Anaerobic Biodegradability	796.3140	not available	not available	not available	Gas production

[a]Requirements to consider a test compound biodegradable under the conditions in the test system. In the Modified Sturm Test, a test compound is considered readily biodegradable if the amount of CO_2 evolved is greater than 60% of the theoretical amount of CO_2 available from the polymer. In addition, greater than 60% must be reached within 10 days after reaching 10% of theoretical CO_2.

[b]Theoretical Oxygen Demand.

[c]Loss of Parent Compound.

[d]>20% loss of DOC implies that the test compound is inherently biodegradable; >70% DOC loss is evidence of ultimate biodegradability.

[e]Considered for regulatory guideline test.

indicates a high potential for biodegradability. The analytical measurement for these methods is usually DOC, as shown in Table 3.3.

Inherent Anaerobic Biodegradability

Anaerobic biotreatment is a major process in wastewater treatment throughout the world, and is found in anaerobic sludge digesters and household septic tanks. As oxygen is not consumed in anaerobic systems, respiration tests based on dissolved oxygen are not applicable. Methane and carbon dioxide gas evolution are the most common measurements of anaerobic activity.

Anaerobic biodegradation tests determine the rate and the degree of anaerobic biodegradation by measuring the evolved volume of CO_2 and methane as a function of time of exposure to anaerobic digester sludge. The evolved volume of gases is measured by pressure readings from sealed containers. The amount of biodegradation is determined by comparing the volume of gases produced in a 60-day test to the amount that theoretically could be produced from the polymer.

A high biodegradability result with a polymer in an anaerobic test system suggests that the polymer will likely be biodegradable in a sewage treatment plant, anaerobic digesters, and natural anaerobic environments such as sediments, flooded soils, and swamps. The validity of gas production as an indicator of anaerobic activity is contingent on the presence of active methane-producing (i.e., methanogenic) bacteria. The test system must be completely protected from exposure to oxygen, which is toxic to anaerobic bacteria-producing methane. Oxygen contamination in the test system also can lead to negative gas pressure due to absorption of oxygen into the test liquid.

Simulation Tests for Aerobic Sewage Treatment

Simulation test methods can be used to measure ultimate biodegradability of a polymer in a model of a WWTP based on activated sludge technology. Biodegradability in these tests is typically reported as percent DOC or percent chemical oxygen demand (COD) removed within a given retention time. These tests are considered confirmatory biodegradation tests for fate in an activated sludge wastewater treatment plant. Although the Coupled Units test is the only broadly known simulation test (OECD, 1981) at the present time, it has fallen out of favor in some circles because of the mixing of test and control unit activated sludge, leading to contamination of the control unit and dilution of the sludge. In addition, synthetic wastewater is used in the test system. Synthetic wastewater can have nutrient deficiencies and different microbial populations compared to actual sewage wastewater. In addition, pH can be difficult to control in Coupled Units tests.

There are alternatives to the Coupled Units test. In both the Porous Pot and the Continuous Activated Sludge (CAS) tests, test and control units are completely separated. Actual wastewater from a municipal wastewater treatment plant is used in these tests. The Porous Pot test method is under final review at the American Society for Testing and Materials (ASTM) and is in the process of becoming a U.S. EPA Office of Pollution Prevention and Toxic Substances (OPPTS) guideline method in the near future. The CAS simulation method has been accepted by U.S. EPA for Premanufacturing Notification (PMN) submissions.

LIMITATIONS OF REGULATORY TEST GUIDELINES

It is important to note that all regulatory biodegradability methods were designed for water-soluble test substances with the possible exception of the inherent anaerobic biodegradability test. Adapting biodegradability methods for virtually water-insoluble polymers can be difficult. Dosing and validation of the test system become troublesome for virtually insoluble polymers (e.g., emulsions). In some cases, the results of fate tests for removal by biodegradability are suspect because of the insoluble nature of some polymers. The total potential for removal by all possible mechanisms, including filtration, precipitation, and sorption, should be understood and reported in all tests for biodegradability.

The low water solubility of some polymers does not usually affect the feasibility of measurement of physical-chemical properties (e.g., solubility) when ^{14}C-labeling is used. However, in regulatory biodegradability testing, measurement endpoints are non-chemical-specific in nature (i.e., DOC, CO_2 evolution, O_2 consumption) and can lack ^{14}C-labeling. DOC measurements can be particularly unreliable in testing of virtually water-insoluble polymers. Evidence of biodegradability may be incorrectly assumed when the limited amount of available DOC is mineralized from the aqueous phase. However, this is a false positive result in terms of overall polymer biodegradability. For polymers with limited solubility, CO_2 evolution or O_2 consumption measurements are more reliable than DOC measurements. In general, synthetic polymers with limited solubility tend not to be readily biodegradable when evaluated with CO_2 evolution or O_2 consumption endpoints. A biodegradable insoluble substance such as cellulose should be used as a positive control for biodegradability tests of limited-solubility polymers.

Because of the low water solubility of some polymers, it has been suggested that biodegradation test concentrations of polymers be increased beyond recommended test concentrations. Most aquatic and wastewater biodegradability methods suggest a test concentration of polymer below 20 mg/L or 20 mg/kg. By adding the polymeric materials well above this level (e.g., up to 1,000 mg/L or 1,000 mg/kg) higher measurement endpoint signals can be generated. If microbial toxicity is not an issue, this is a reasonable alternative to chemical-specific analysis or ^{14}C-labeling.

OVERVIEW OF AUXILIARY TEST METHODS

Aquatic and Wastewater

Auxiliary test methods have been developed to answer questions that are not yet adequately covered with regulatory guideline tests. These auxiliary test methods have been developed in industrial laboratories (e.g., Unilever and the Procter & Gamble) and have been used by contract laboratories (e.g., Roy F. Weston, Inc., 1977–1995). New fate tests that are in the process of being developed will more fully describe the behavior of a chemical in both aerobic and anaerobic systems. These tests will determine the amount of ultimate biodegradation, as do many of the regulatory guideline tests, but also will determine the rate of disappearance of the parent polymer and appearance of metabolite(s).

Microbial Inhibition

An important auxiliary test is the microbial inhibition (toxicity) test. It is important to emphasize the objectives of biodegradation tests and microbial toxicity tests. The objective of biodegradation testing is to determine the ability of microorganisms to metabolize or mineralize a chemical (i.e., polymer) at relevant exposure concentrations. The objective of microbial toxicity testing is to determine the concentration at which a chemical impairs the ability of wastewater treatment microorganisms to metabolize. The inexpensive microbial inhibition test can prevent the erroneous negative result of a biodegradation test. Microbial inhibition could cause a company to mistakenly discard a polymer that "was not biodegradable" when in fact no biodegradation was possible because the microorganisms were inhibited at test concentrations.

The only test that currently addresses wastewater treatment microorganism toxicity in U.S. or European regulatory guidelines is OECD Method 209, Activated Sludge, Respiration Inhibition Test (OECD, 1984). This test is a screening tool to assess the impact of a chemical on the overall respiration rate of activated sludge. It determines microbial inhibition based on dissolved oxygen measurements at activated sludge concentrations found in WWTP aeration basins. Another general test for microbial inhibition is based on the measurement of BOD (Marks, 1973). This test is for low microorganism concentrations such as those found in rivers and lakes. A more specific test for measuring microbial toxicity in high strength inoculum is available from contract laboratories (e.g., Roy F. Weston, Inc., 1977–1996). In this test, a known amount of ^{14}C-labeled glucose is added to an inoculum. The change in $^{14}CO_2$ evolution from the test inoculum is measured after a specific time period for each polymer concentration added. The ability of a mixed microbial population to metabolize biodegradable organic chemicals (e.g., glucose) is a measure of heterotrophic activity (HA). Inhibition of HA by addition of a toxicant is a measure of the population's reduced metabolic activity. A 50% reduction in the heterotrophic activity (HA_{50}) is defined as the calculated concentration of polymer yielding a 50% reduction in ^{14}C-labeled glucose uptake.

Modified Closed Bottle Test

The standard Closed Bottle test is based on measurement of BOD where oxygen consumption is measured from a virtually airtight single liquid phase containing inoculum (i.e., single-phase system). Practical difficulties in the single-phase system include the tendency of insoluble polymers to float in the test inoculum medium and the consequent problem of sample loss when inoculum bottles are filled completely. Further, some polymers solids will adhere to the sides of test bottles in the absence of effective agitation. To counter this problem, a two-phase system was devised in which an air phase was left above the liquid inoculum phase (van der Zee et al., 1994). This system allows more effective mixing and greater availability of oxygen for biodegradation. Compared to the single-phase system, faster times to completion of biodegradation as well as higher precision (i.e., lower standard deviation) of the BOD measurements have been found.

Batch Activated Sludge Test (BAS)

Typical biodegradability tests that measure ultimate biodegradation use a very low concentration of inoculum to simulate surface waters. An inherent biodegradability test to measure ultimate biodegradation by CO_2 evolution from a high strength inoculum is available. The inoculum concentration of activated sludge is comparable to the sludge concentration in an aeration basin of a secondary WWTP. The BAS test requires radiolabeled materials because of the high background of unlabeled CO_2 from the activated sludge. The ^{14}C-labeled test substance is put into a flask containing one liter of activated sludge of approximately 2,500 mg/L total suspended solids (TSS). Test flasks are fed with synthetic feed at each sampling period. The $^{14}CO_2$ evolved from the test system is measured by liquid scintillation counting (LSC). The sludge filtrate also is sampled and analyzed by LSC. The total amount of radioactivity in the volatile (e.g., $^{14}CO_2$), soluble, and insoluble fractions is used to report the ^{14}C-label mass balance.

Continuous Activated Sludge (CAS)

The CAS test was developed because the OECD Coupled Units test was deemed by some to be unrealistic. This test simulates a secondary WWTP by using a mixing chamber, an aeration basin, and a clarifier. The sludge biomass solids can be wasted (i.e., removed from the system) or recycled to control the sludge concentration and the sludge retention time. Wastewater flow can be adjusted to control the hydraulic retention time. Operational measurements such as BOD, COD, pH, TSS, nitrogen, and phosphorus are routinely taken. After the acclimation phase, the CAS system is sampled for polymer. Both the liquid and the solids are sampled. Both sorption and biodegradation potential of the polymer in an activated sludge matrix can be determined.

Porous Pot Test

The Porous Pot Test simulates an aeration basin at a secondary WWTP. Both sludge retention time (SRT) and hydraulic retention time (HRT) can be monitored and controlled. Typical measurements include BOD, COD, pH, TSS, nitrogen, and phosphorus. Both removal by sorption and biodegradation information can be obtained by this test.

Anaerobic Heterotrophic Activity Test: Septic Tank Bacterial Toxicity

This test is designed to determine the bacterial toxicity of the polymer to a mixed microbial population in septic tank media (i.e., septage), as measured by ^{14}C-glucose uptake. This is an important test for measuring the impact on septic tank performance from products. A high volume of products in wastewater are disposed to septic tank systems (e.g., approximately 25% of all U.S. households use septic tank systems). If a polymer is toxic to septage microorganisms, mechanical breakdown of the septic tank system could occur. No current regulatory guideline test addresses this issue.

Septage is collected from a functioning septic tank receiving toilet and "gray" (e.g., kitchen and laundry) water. There are both aerobic and anaerobic pockets in a septic tank; aerobic microorganisms as well as facultative and methanogenic anaerobic microorganisms are present in the tank. The septage is incubated with a series of concentrations of the polymer for several hours. A known quantity of ^{14}C glucose then is added and incubated for an additional few hours. The reduction in the amount of ^{14}C removed from solution due to added polymer is measured by LSC.

Batch Settling Test: Effect on Settling of Suspended Solids

Many chemicals are disposed to the sewer and are treated by municipal sewage treatment plants. This test can determine the effect of polymers on the settling of solids in domestic wastewater as measured by the accumulation of solids in an Imhoff cone. Settling of solids is important in the purification of wastewater. In this test, a wastewater sample is collected from the influent of a municipal sewage treatment plant, TSS measurement is taken, and the wastewater is added to an Imhoff cone. The polymer is added, and the volume of solids that can settle is recorded after set time periods and compared to both positive (e.g., ferric chloride) and negative (e.g., alkylbenzene sulfonate) controls. Two polymer concentrations typically are used.

METHODOLOGICAL ISSUES

Inoculum Samples

One common assumption in taking inoculum samples from single WWTP facilities for aerobic biodegradation testing is that a similar inoculum could be obtained elsewhere. In fact, this assumption is a basis for most of the regulatory testing guideline methodologies for aquatic biodegradation testing around the world. Some groups are beginning to look at this assumption to see if, and under what circumstances, it is valid. European WWTP facilities are usually run at significantly higher activated sludge suspended solids concentrations and shorter sludge retention times than WWTP facilities in the United States. Therefore, European WWTP inoculum characteristics may not always be assumed to be the same as U.S. WWTP inoculum characteristics.

Limited Solubility

A major methodological issue involves dosing methods for low solubility polymers into biodegradability test systems. The ideal manner for dosing a low solubility polymer is to create a suspension of the polymer in the test system; this maximizes the surface area of the polymer for the inoculum and thus provides the greatest potential for biodegradation. Some options are the following:

- Sonification and/or use of a blender or similar device (e.g., Polytron™) to emulsify the polymer in the test media.
- Use of a volatile organic solvent as a carrier.
- Adding the polymer to a glass wool, glass fiber filter, watch glass, or other inert material and then putting the inert material containing the polymer into the test media.
- Use of a dispersing or solubilizing agent (e.g., linear alkylbenzene sulfonate) to add the polymer to the test system.
- Adding the insoluble material by direct weight, which is the simplest and probably least effective dosing method.

Additional discussion of dosing methods was provided in Chapter 2. Each of the above methods has disadvantages that may compromise the accuracy and the precision of the dosing methodology compared to the typical stock solution preparation for a soluble chemical. Further, it should be noted that overdilution of the dispersant phase of polymer emulsions will often destabilize dispersed polymer, causing it to form large polymer particles. These larger particle sizes are highly susceptible to coagulation. Filtration of solids is a common industrial method for treating industrial wastewater that contains water-based polymer emulsions and other dispersed polymers. Therefore, the potential for removal of polymers by filtration of solids must be understood in fate tests with dispersed polymers.

Use of Radiolabeled Materials

When practical, the use of ^{14}C-radiolabeled test materials is highly recommended. However, it may be cost-prohibitive to label certain types of polymeric materials. In addition, ^{14}C-radiolabeling will not confirm the chemical identity of polymer analytes. Restrictions on transport and waste disposal of radioactive materials makes it difficult to work with radiolabeled materials. Even with these constraints, there are certain advantages in carrying out ecological fate testing with radiolabeled test substances. Because there is very low background in analyzing for ^{14}C by LSC, test concentrations can be much lower than in nonlabeled studies. For example, whereas the testing concentration in an unlabeled study is often constrained to approximately 10 mg/L as polymer, the testing concentration for a ^{14}C-labeled chemical can be three orders of magnitude lower, or 0.01 mg/L as polymer. This lower level may be much more representative of actual environmental concentrations than the higher level. Further, if polymer-related toxicity is possible, then the use of ^{14}C-labeled polymers usually allows testing at nontoxic polymer concentrations. Further, because ^{14}C-label can be followed with LSC, mass-balance for the polymer can be obtained during the biodegradability test; that is, the ^{14}C-label can be traced in the form of $^{14}CO_2$ evolved, the amount remaining in the aqueous phase, and the amount sorbed to any solids in the test system. Metabolite identification and concentrations along with parent concentrations can be monitored with the use of analytical techniques such as thin layer chromatography (TLC) or high performance liquid chromatography (HPLC) with a radiochemical detector. In addition, with the use of multiple radiolabels, information about different subunits of polymers can be monitored. For example, if one label has been put on a repeating subunit and another on the backbone of the polymer, degradation of each part of the polymer can be traced. If multiple types of radiolabels are used (e.g., ^{14}C and ^{3}H), information can be obtained on mechanisms of biodegradation.

As more definitive fate methodologies become available for ^{14}C-labeled materials, a stronger case for developing a radiolabeled test compound can be established. The cost of synthesizing the radiolabeled material may be offset by the amount of information that can be obtained from fate tests using the radiolabeled material.

Chemical-Specific Analysis

The cost of chemical-specific analysis of "difficult" materials such as polymers can be very expensive. However, the need for chemical-specific analysis is evident when non-chemical-specific analyses do not give reliable results.

Analytical techniques that are theoretically amenable to polymeric materials are gel permeation chromatography (GPC) and capillary zone electrophoresis (CZE), as well as the more traditional methods of HPLC, TLC, and gas chromatography (GC). The typical detection methods for these analytical techniques are non-chemical-specific (e.g., UV–visible, electrochemical, fluorescence, and radiochemical). Therefore, analytical standards are required to identify the polymer analytes. Mass spectrometry (MS) may allow identification of polymer analytes for which no standards have been synthesized. However, the successful combination of chromatography with MS has been very difficult for many chemicals, including polymers. GC/MS took decades of experimentation before it became widely used and accepted. The only combined liquid chromatographic (LC) technique that is widely available is LC/MS. Although this method has been experimented with for well over a decade, LC/MS is still not widely used by the general scientific community. CZE/MS is beginning to be evaluated as a technique for separating and identifying high molecular weight polymers.

The following is a summary of relatively new analytical techniques that show promise for use with polymers, but the high cost of these techniques must be emphasized. Regulatory programs considering adoption of any analytical technique to generate fate data for new polymers should seriously balance analytical costs with actual need for polymer-specific fate information. Additional discussion of analytical techniques for polymers is found in Chapters 4 and 5. New interfaces for LC/MS, such as atmospheric pressure ionization (API-LC/MS), and electrospray (ES-LC/MS) show promise for the analysis of polymers

Gel Permeation Chromatography

In GPC, a gel allows the entry of smaller-size molecules while excluding larger ones. Smaller polymer fragments will elute more slowly in GPC gels, thereby separating themselves from the larger fragments. The technique has been used with some success to characterize the metabolites of plastics. In addition, OECD has drafted guidelines for the determination of the molecular weight and molecular weight distribution of polymers using GPC (OECD, 1995a, b) along with a draft guideline on solution-extraction behavior of polymers in water (OECD, 1995c).

The limited solubility of some polymers requires the use of organic solvents and relatively harsh conditions. When unanticipated polymer degradation during analysis results in polymer cross-linking and decreased solubility, GPC results will not be representative of actual degradation potential in the environment. Like other expensive techniques, GPC should be used more for in-depth analysis of polymer fate than as a screening tool.

Capillary Zone Electrophoresis

Electrophoresis has been used for many years to separate DNA material, but CZE has been successfully used to separate large molecules on the basis of charge density only for the past ten years or so. One of the most distinctive properties of CZE is electroosmotic flow. This bulk movement of solvent is caused by the small charge (zeta) potential at the silica/water interface, which induces the adsorption of a minute excess of anionic species in the static diffuse double layer. This produces an excess of cationic species in bulk solution, which migrate toward the cathode, producing a pluglike flow that has a flat velocity distribution across the capillary diameter, deviating only within a few nanometers of the capillary surface. This produces the extremely high resolution that is possible with this technique, much higher than other chromatographic methods.

QUALITY ASSURANCE AND CONTROL

Quality assurance and control (QA/QC) can follow Good Laboratory Practice (GLP) regulations based on OECD GLP principles and guidelines (OECD, 1992a–e; 1993b,c; 1994). QA/QC programs form the basis for documentation of data reliability, and GLP procedures are typically required when ecological fate data are submitted to regulatory agencies.

REFERENCES

European Union. 1992. Part C: Method for the Determination of Ecotoxicity, in *Official Journal of the European Communities,* No. L 383A. Brussels, Belgium: European Union.

Marks, P. J. 1973. Microbiological Inhibition Testing Procedure, in *Biological Methods for the Assessment of Water Quality.* ASTM STP 528. Philadelphia, PA: American Society for Testing and Materials.

OECD. 1981. *OECD Guidelines for Testing of Chemicals.* Section 3 — Degradation and Accumulation, #303A — Simulation Test — Aerobic Sewage Treatment: Coupled Units Test. Paris, France: Organisation for Economic Co-operation and Development.

OECD. 1984. *OECD Guidelines for Testing of Chemicals.* Section 2 — Effects on Biotic Systems, #209 — Activated Sludge, Respiration Inhibition Test. Paris, France: Organisation for Economic Co-operation and Development.

OECD. 1992a. *Principles of Good Laboratory Practice.* OECD Environment Monograph No. 45. Paris, France: Organisation for Economic Co-operation and Development.

OECD. 1992b. *Guides for Compliance Monitoring Procedures for Good Laboratory Practice.* OECD Monograph No. 46. Paris, France: Organisation for Economic Co-operation and Development.

OECD. 1992c. *Guidance for the Conduct of Laboratory Inspections and Study Audits.* Environment Monograph No. 47. Paris, France: Organisation for Economic Co-operation and Development.

OECD. 1992d. *Quality Assurance and GLP.* Environmental Monograph No. 48. Paris, France: Organisation for Economic Co-operation and Development.

OECD. 1992e. *Compliance of Laboratory Suppliers with GLP Principles.* Environment Monograph No. 49. Paris, France: Organisation for Economic Co-operation and Development.

OECD. 1993a. *OECD Guidelines for Testing of Chemicals.* Section 1—Physical-Chemical Properties and Section 3—Degradation and Accumulation. Paris, France: Organisation for Economic Co-operation and Development.

OECD. 1993b. *The Application of the GLP Principles to Short-Term Studies.* Environment Monograph No. 73. Paris, France: Organisation for Economic Co-operation and Development.

OECD. 1993c. *The Role and Responsibilities of the Study Director in GLP Studies.* Environment Monograph No. 74. Paris, France: Organisation for Economic Co-operation and Development.

OECD. 1994. *The Application of the GLP Principles to Field Studies.* Environment Monograph No. 50. Paris, France: Organisation for Economic Co-operation and Development.

OECD. 1995a. *Draft Guideline: Determination of the Number-Average Molecular Weight and the Molecular Weight Distribution of Polymers Using Gel Permeation Chromatography.* Paris, France: Organisation for Economic Co-operation and Development.

OECD. 1995b. *Draft Guideline: Determination of the Low Molecular Weight Content of a Polymer Using Gel Permeation Chromatography.* Paris, France: Organisation for Economic Co-operation and Development.

OECD. 1995c. *Draft Guideline: Solution-Extraction Behavior of Polymers in Water.* Paris, France: Organisation for Economic Co-Operation and Development.

Roy F. Weston, Inc. 1977–1996. *Ecological Fate and Sludge/Bacteria Inhibition Test Protocols.* Lionville, PA: Roy F. Weston, Inc.

U.S. EPA. 1994. *Code of Federal Regulations.* Title 40—Protection of Environment, Chapter 1—EPA, Subchapter R—TSCA, Part 796—Chemical Fate Testing Guidelines. 40 CFR 796. Washington, DC: U.S. Government Printing Office.

U.S. FDA. 1987. *Environmental Assessment Technical Assistance Handbook.* FDA/CFSAN-87/30. Washington, DC: U.S. Food and Drug Administration.

van der Zee, M., L. Sijtsma, G. B. Tan, H. Tournois, and W. de Wit. 1994. Assessment of biodegradation of water insoluble polymeric materials in aerobic and anaerobic aquatic environments. Chemosphere 28:1757–71.

Chapter *4*

Biodegradation Testing of Polymers in Soil

■ RICHARD BARTHA AND ASHA V. YABANNAVAR
Rutgers University

INTRODUCTION

The soil environment is virtually unsurpassed in its diversity of catabolic (i.e., degradative) microorganisms (Bartha, 1990). Biodegradation rates higher than those in soil can be achieved in intensely aerated liquid and solid phase reactors (e.g., activated sludge, fermenters, composting), but no single environment can achieve the degradation of such a range of natural and xenobiotic chemicals as soil.

Besides its microbial diversity, soil has additional advantages as a catabolic environment. Soil is commonly aerobic with anaerobic microsites within aggregates. Soil organic matter (humus) is a ready source of micronutrients and cosubstrates. The high absorption capacity of clays and humus mitigates the toxic effects of some substrates and their metabolites. These characteristics add up to an exceptionally favorable environment for biodegradation. It can be said that if a chemical is inherently biodegradable, it will be degraded in soil. A chemical that fails to be degraded in biologically active soil is likely to be recalcitrant (i.e., resistant to biodegradation) in other natural environments. This wisdom is reflected in the fact that tests for inherent biodegradability are commonly conducted in a soil matrix. As usual, some exceptions can be found to these generalizations. The dechlorination of some haloaromatics occurs at low redox potentials that are not attained in vadose soil (Kuhn and Suflita, 1989). Also, the covalent binding of some reactive chemicals or their degradation intermediates to soil humus can retard the ultimate degradation of these chemicals as compared to humus-free environments (Bartha et al., 1983).

53

The soil environment poses a substantial challenge to measurement of biodegradation of the vast majority of organic chemicals, including polymers. The advantages of soil as a matrix for biodegradation tests are balanced by the interference of this matrix with numerous monitoring and analytical techniques. Soil is a sticky, nontransparent medium. For example, it interferes with the visual monitoring and retrieval of exposed polymer samples and with most of the traditional substrate-depletion/biomass-increase assays applicable to liquid media. Soil is also a highly variable entity in terms of its composition and biological activity, which vary with location and season, pretreatment, and storage. To date, it has been impossible to standardize soil as a biodegradation test medium. Most test protocols specify the use of three soils with different texture characteristics (i.e., sandy, loam, and clay) textured soil. Some of the early recommendations concerning the collection, handling, and storage of soils intended for use in biodegradation tests (Pramer and Bartha, 1972) were incorporated into these protocols. Some of the variability inherent in natural soils is addressed by the use of positive controls in biodegradability tests.

FEATURES OF AN "IDEAL" BIODEGRADATION MONITORING TECHNIQUE

As this chapter will discuss specific shortcomings of various monitoring techniques for polymer biodegradation in soil, it may be useful to define what would constitute an "ideal" monitoring technique. Obviously, such a technique does not exist at present and most likely never will. However, because monitoring techniques are still very much in a state of flux and development, at least certain features of an "ideal" technique have a chance to be realized.

Perhaps most important, a biodegradation test in soil should be predictive. The test should give a close approximation of what happens to the material when it enters the soil environment in the field. As it is impractical to test for all climatic variables, biodegradation in soil is measured under near-optimal conditions in terms of moisture, aeration, temperature, and pH (Pramer and Bartha, 1972). It is (or should be) understood that biodegradation in soil may be reduced or prevented by excessive dryness, anoxic conditions, freezing temperatures, or pH extremes. In addition, the ideal biodegradability screening technique would be broadly applicable to various test materials and would be easy to standardize. It should not require a separate set of analytical techniques for every test material. The monitoring should be nondestructive of the test samples, allowing time-course measurements without excessive numbers of replicates. Finally, the test should be sensitive, accurate, and reasonable in cost of supplies, instrumentation, and personnel. Interpretation of the test results should be simple and unambiguous.

Although the ideal biodegradation screening technique does not exist currently, the monitoring of carbon-14–labeled carbon dioxide ($^{14}CO_2$) evolution from ^{14}C-labeled test substrates comes reasonably close to it. It is simple, nondestructive to the test substrate, and sensitive. Unfortunately, it requires the expensive and often technically challenging custom synthesis of radiolabeled test substrates. Acquisition of labeled test material is especially difficult in the case of polymers and composite materials. For these reasons, only a few studies have been published on the biodegradation of ^{14}C-labeled polymers (Albertsson, 1978, 1980; Sielicki et al., 1978); so the critical examination of other test methods is necessary. Some of these methods were reviewed by Aminabhavi and others (1990). The following sections update and expand this review with a special focus on the soil environment.

ESTABLISHED METHODS TO MEASURE POLYMER BIODEGRADATION IN SOIL

Designating certain measurements as "established" is somewhat arbitrary. For the sake of this discussion, "established" methods have been used for well over a decade. Only the following four broadly applicable monitoring techniques are appropriate for virtually any polymer. Case studies of measuring biodegradation plastics in soil are provided in Chapter 9.

Carbon Dioxide Evolution

The broad applicability, simplicity, and nondestructive nature of the carbon dioxide evolution test makes it rather attractive. As indicated in Chapter 3, it is the basis of standardized biodegradability tests specified by the U.S. Environmental Protection Agency (U.S. EPA), the U.S. Food and Drug Administration (U.S. FDA), and the Organisation for Economic Co-operation and Development (OECD). If ^{14}C-labeled material is available for the test, it is important for the label to be either uniform or located in the "difficult" portion of the molecule; otherwise the test may give misleading results (Atlas and Bartha, 1993).

The radiolabel allows sensitive and specific measurement of substrate-derived CO_2 at low substrate concentrations. As some of the radiocarbon is incorporated into biomass and humic chemicals, a $^{14}CO_2$ evolution in excess of 50% of the added radiocarbon is interpreted as complete or near-complete biodegradation. The specified test protocols recommend the use of biometer flasks (Bartha and Pramer, 1965) for these tests. This is somewhat curious, as this apparatus (Figure 4-1) was developed for the accurate measurement of unlabeled CO_2 evolution from soil, and its features for keeping out atmospheric CO_2 are nonessential for ^{14}C-based measurements. Nevertheless, the flask seems to be favored for its compact design. An essential precaution is to ascertain that radioactivity trapped in the alkali of the sidearm is $^{14}CO_2$ rather than some other acidic and volatile material. This is easily accomplished by barium chloride ($BaCl_2$) precipitation of the carbonate followed by a recount of the supernatant. All $^{14}CO_2$-derived radioactivity should be removed by the precipitation.

If radiolabeled test material is not available or prohibitively expensive to synthesize, net CO_2 evolution from the test substrate can be measured.

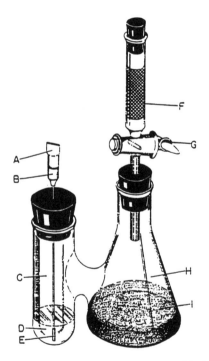

Figure 4.1. Biometer flask for measuring the ultimate biodegradation (mineralization) of xenobiotic substances in soil. CO_2 evolved from soil plus xenobiotic substance (I) in the main flask (H) is trapped in the alkali (D) of the side arm (C). The alkali is periodically retrieved, and the trapped CO_2 is measured by volumetric titration.

This measurement assumes that the addition of the test substrate does not change the background CO_2 evolution of the soil, and thus net CO_2 from the test substrates equals test minus soil blank. A recent examination of this assumption showed it to be correct with only a few exceptions (Shen, 1994). Glucose, frequently used as a positive control in biodegradation tests, at some concentrations stimulates soil organic matter degradation, inflating the apparent net CO_2 evolution (Sharabi and Bartha, 1993). Why glucose acts in this manner is not clear; diverse natural and xenobiotic substrates tested have failed to produce a priming effect comparable to that of glucose (Sharabi and Bartha, 1993; Shen, 1994). Therefore, it appears that net CO_2 evolution measurements give generally valid results. For use as a positive control, benzoate can be recommended instead of glucose.

Because in net CO_2 evolution measurements the ratio of test and background CO_2 is critical for the accuracy of the measurement, good results are obtained when the test substrate is applied at 400 to 500 mg/kg. Work with 200 mg/kg, as suggested by U.S. FDA, U.S. EPA, and OECD protocols, requires a very precise technique, and net CO_2 measurements should not be attempted at concentrations less than 100 mg/kg. With substrates that are expected to be degraded very slowly or incompletely, as is the case with some synthetic polymers, test substrate concentrations as high as 1% (w/w) have been used (Yabannavar and Bartha, 1993, 1994).

Just as in the case of $^{14}CO_2$ evolution, net CO_2 evolution is related to the maximum theoretical CO_2 evolution to be expected from the test substrate. CO_2 evolution in excess of 50% of theoretical is interpreted as complete or near-complete biodegradation. For such calculations, the tester does not need to know the composition of the polymer but only needs its total carbon content, which is an inexpensive analysis. For testing proprietary materials, this can be a considerable advantage.

In aerobically incubated soil, oxygen consumption is the mirror image of CO_2 production. In principle, net CO_2 production measurements could be replaced by net oxygen consumption measurements. Numerous devices have been constructed to measure oxygen consumption by soil. However, oxygen consumption measurements in experiments lasting from several weeks to several months, have proved to be less reliable and less accurate than CO_2 evolution measurements. This situation may change with the relatively recent introduction of advanced computerized respirometers using gas sensors or electrolysis of water. These devices are quite expensive but appear to be suitable for accurate long-term measurements of oxygen consumption by soil samples. To date, we are not aware of polymer biodegradation tests in soil monitored by oxygen consumption. Polycaprolactone degradation by pure and mixed cultures have been monitored by oxygen consumption measurements (Lefebre et al., 1994), and measurements in soil or compost may follow.

Weight Loss Measurement

The conversion of a solid polymer to CO_2, water, and other nonsolid products results in weight loss. The measurement of weight loss is simple, broadly applicable, and only requires an analytical balance as instrumentation. However, some of the weight loss can be masked by the incorporation of oxygen into the solid. More important, it is both tedious and inaccurate to free partially degraded polymer from soil particles and biomass. Conservative cleaning tends to be inadequate, whereas more aggressive cleaning leads to fragmentation and material loss. Depending on the size of the polymer particles and their subsequent degradation and fragmentation, the retrieval of the polymer sample incubated in soil also may be rather tedious and inaccurate. For these reasons, weight loss measurement popular in early studies (Booth and Robb, 1968; Klausmeier, 1972; Wendt et al., 1970) has been used in recent soil studies (Goheen and Wool, 1991; Mergaert et al., 1993; Yabannavar and Bartha, 1993, 1994) only in combination with other techniques.

To make residual weight measurements more accurate and convenient, Yabannavar and Bartha (1993, 1994) have applied solvent extraction techniques to the retrieval of polymers incubated in soil. Such extractions are used routinely for pesticide, hydrocarbon, and many other hydrophobic contaminants. The extracted polymer residue was weighed after solvent removal by precipitation/filtration and/or solvent evaporation. Although this approach worked well for PVC-based plastics, it was not applicable to materials that contained solvent-insoluble additives. Moreover, some polyethylene-based materials that were amenable to solvent extraction in their native state became partially insoluble during degradation because of the cross-linking of polymer strands.

Viscosity Measurements

If polymer residues can be extracted from soil into solvent solution, the viscosity of these solutions is an indication of polymer molecular size. Comparing the viscosity that a polymer lends to its solvent solution prior to and after exposure in soil will indicate whether any changes have occurred in the size of the polymer molecules during the incubation. As long as a suitable solvent is available for the polymer and standardized conditions of concentration and temperature are used, this approach is broadly applicable. Moreover, standard procedures have been developed for viscosimetric examination of various polymers. Yabannavar and Bartha (1993) adapted an American Society for Testing and Materials (ASTM) standard procedure for viscosimetric examination of PVC samples in cyclohexane solution to test for changes in molecular size during 5 weeks of exposure in soil The measurements were quite reproducible and relatively simple, but in the described case no significant changes in PVC molecular size occurred during exposure in soil. Although potentially useful, viscosimetry has been rendered essentially obsolete by newer gel permeation chromatographic (GPC) techniques (see below) that are more sensitive and provide much more information than the viscosimetric measurements.

Loss of Tensile Strength (Elongation)

The loss of tensile strength can be an early and sensitive indicator of polymer deterioration/degradation in soil. Degradation that is barely perceptible as weight loss can cause large changes in tensile strength. The test is broadly applicable to film-type polymer materials and has a long history of use. In the United States, it is performed according to standard method ASTM D882-83. The drawback of this test approach is that it can be somewhat misleading in terms of environmental fate. The mechanical properties are disproportionately affected by changes in the additives to the polymers. A 90% decrease in tensile strength (elongation) may be brought about by the mineralization of 5% of the total polymer carbon. The test is also somewhat unpredictable in that in some films deterioration/degradation actually increases elongation without breakage, which is the opposite of the expected outcome.

Tensile strength changes are better indicators of polymer deterioration than of ultimate degradation. Because of the importance of mechanical properties, there is little doubt that the use of tensile strength tests will continue. These tests are also good indicators of moderate photochemical damage that is difficult to detect by other means.

LIMITATIONS OF ESTABLISHED MONITORING METHODS

Of the four broadly applicable monitoring methods described above, loss of tensile strength is least correlated with ultimate biodegradation. Loss of residual weight is more relevant in this respect, but it is a tedious and relatively inaccurate method. Attempts to make residual weight measurements more accurate and less laborious through solvent extraction (Yabannavar and Bartha, 1993, 1994) were successful only for certain types of polymers. The need to find a suitable solvent for each type of polymer and the need to establish recovery efficiencies not only for the starting material but for the partially degraded product complicate the use of this technique. Aside from the methodological difficulties, weight loss of composite materials may reflect the degradation of additives and not of the polymer. Only weight loss well in excess of the total amount of additives is indicative of polymer degradation, provided that the retrieval and the measurements are accurate. In this respect, solvent extraction, where applicable, offers certain advantages. As the polymer is precipitated, dried, and weighed, dissolved additives may be quantified in the solvent extract by other means (e.g., gas chromatography). In this way, weight losses can be specifically attributed to the polymer or the additives (Yabannavar and Bartha, 1993). The solvent-recovered polymer can be also subjected to viscosimetric analysis or the more powerful GPC technique. Tensile strength, residual weight, and viscosimetric measurements are all destructive analyses, and replicate samples need to be analyzed for each time point.

Measurements of $^{14}CO_2$ and net CO_2 evolution are both closely related to ultimate degradation of the test material. They are nondestructive tests, suitable for time-course monitoring of the same samples. Measurement of $^{14}CO_2$ evolution offers high sensitivity and an assurance that changes in background CO_2 evolution will not influence the test results. These advantages are balanced by the high cost and technical difficulty of radiolabeled-sample procurement. Licensing requirements and radioactive waste disposal are additional inconveniences.

In using radiolabeled test materials, analytical sensitivity allows very low concentrations. Nevertheless, one is well advised not to go below the 200 mg/kg (i.e., 0.2 mg/g) concentration recommended by the U.S. FDA, U.S. EPA, and OECD for soil biodegradability tests unless the material is known to be extremely toxic. Test materials at too low concentrations may give rise to false negatives if sorption to organics and/or inorganics renders them unavailable, or if they fail to induce degradative enzymes below a certain threshold concentration (Alexander, 1994). Sharabi and Bartha (1993) found that ^{14}C glucose at 0.002 to 0.004 mg/g was converted to $^{14}CO_2$ to a lesser extent than at 1 mg/g. In this case, follow-up investigations showed that ^{14}C glucose at concentrations ranging from the 0.002 mg/g to 0.020 mg/g levels was used by microorganisms with higher metabolic efficiency than at 0.2 mg/g and above. In other words, at very low glucose concentrations a larger proportion of the ^{14}C was incorporated into cell biomass than into $^{14}CO_2$, whereas at higher glucose concentrations the reverse occurred (Shen, 1994). This is an additional reason why too low test material concentrations may underestimate the true potential for biodegradation.

Net CO_2 evolution requires higher test substrate concentrations than does $^{14}CO_2$ for reliable results. Recent tests indicate that net CO_2 evolution measurements are generally reliable and a priming effect (increase of background CO_2 evolution) caused by certain levels of glucose additions represents exceptional cases. However, in testing of unlabeled composite materials, there is no assurance that the CO_2 is evolved evenly from the various components, which can complicate the interpretation of test results. One normally assumes that for every carbon evolved as CO_2, another carbon is incorporated into microbial biomass. However, this is a very rough approximation, and the CO_2/biomass ratio can vary a great deal with the nature of the substrate, its concentration, and the prevailing environmental conditions. Therefore, if a composite consists of 80% polymer and 20% additive-carbon, a release of 15% carbon as CO_2 may or may not indicate mineralization of some of the polymer material. In such cases, it is better to err on the conservative side and claim polymer degradation only if carbon well in excess of the 20% additive is converted to CO_2.

One may conclude that although $^{14}CO_2$ and net CO_2 evolution are the most useful of the established techniques, ^{14}C-labeled material is expensive and at times unobtainable, and the interpretation of net CO_2 evolution results is rather complicated and occasionally ambiguous.

USE OF GEL PERMEATION CHROMATOGRAPHY

During the past decade, GPC has been emerging as the most broadly applicable and most powerful chemical-specific method for evaluating polymer degradation. ASTM definitions of polymer degradation currently under development rely heavily on this analysis. GPC instrumentation with hot solvent capability, a necessity for the analysis of several polymers, is quite costly. When a polymer degradation process results in cross-linking and decreased solubility, GPC analyses may not yield representative results. GPC and more specialized analyses described in this section are likely to be used more for confirmation and in-depth analysis of polymer fate than as first-tier screening tools.

The desirable properties of polymers are due to long, more or less linear molecules with molecular weight typically in excess of 100,000. True degradation of a polymer, whether by biochemical, chemical, or photochemical means, shortens these polymeric molecules. Degradation may start from either end of the molecule or, more typically, will break up the polymeric molecule into shorter pieces. In GPC, suitable gels will allow the entry of certain smaller-size molecules, while excluding larger ones. These excluded larger molecules will travel faster during column elution in the void volume, whereas smaller molecules will elute more slowly with the bed volume. The technique originally developed for separation and seizing of hydrophilic macromolecules was soon adapted to molecules soluble only in organic solvents. The difficult dissolution characteristics of some plastic chemicals require harsh conditions. Polyethylene is typically chromatographed in trichlorobenzene at 130°C. Any insoluble material needs to be removed for successful chromatography. Because of cross-linking, the solubility of polymers may decrease after partial degradation, especially when photochemical or free-radical reactions are involved. It must be understood that such insoluble molecules will be missed by the GPC analysis process.

The properties of a plastic material are determined largely by its molecular weight distribution. The polymerization process results in a mixture of molecular sizes (polydispersity). To characterize a polymer, its average molecular weight can be calculated in several ways. The following terms are commonly used in characterizing the molecular weight distribution of a polymer:

- Number average molecular weight (M_n) is defined as the total weight of all the molecules in a polymer sample divided by the total number of moles present.
- Weight average molecular weight (M_w) is defined as the mean of the weight distribution of molecular weights.
- Polydispersity (Pd) is defined as the breadth of the distribution of molecular weights in a polymer (M_w/M_n). A monodisperse polymer (uniform molecular size) would have a Pd of 1.0. In a typical polymer, Pd has values of 2 to 5.

In a polydisperse polymer, M_w is biased toward larger and M_n toward smaller molecules. Because the properties of the polymer usually depend disproportionately on the large molecules, M_w is generally more useful than M_n. Determination of both parameters allows the calculation of polydispersity (Pd = M_w/M_n), an important parameter for characterizing the polymer.

All degradation processes would be expected to result in a decrease of M_n and M_w and an increase of Pd. Photodegradation, causing breaks in the longer polymer strands, would have a greater effect on M_w than on M_n. Biodegradation, if biased toward the smaller polymer molecules, would affect M_n more strongly than M_w. Free radical reactions in photo- and autoxidation may result not only in breaks in polymer strands but also in some cross-linking of the strands. Such cross-linking side reactions increase M_w and M_n and thus may to some degree mask the true degradative damage. Cross-linking is likely to interfere with biodegradation and may also decrease solubility in certain solvents. Cross-linking with its consequent solubility problems is the major limitation of the GPC approach to polymer biodegradation measurements. However, GPC measurements were used with generally satisfactory results to measure biodegradation of synthetic polymers in soil (Yabannavar and Bartha, 1994), in compost (Johnson et al., 1993), and by microbial cultures (Benedict et al. 1983a, b; Lee et al., 1991). In the case of polyethylene, pretreatment by heat or irradiation was necessary to prime the materials for subsequent degradation in soil or by microbial cultures.

OVERVIEW OF AUXILIARY METHODS

The previously discussed methods were broadly applicable to various types of polymers. The measurements listed here are applicable only to certain polymeric materials, or not even to the polymer itself but to additives of the formulation. Such determinations help to clarify the overall fate of the composite test material and help to interpret weight loss data by clarifying whether the weight loss was due to polymer or to additive degradation.

Chloride Release

Chlorine is rare in natural organic compounds but is a frequent constituent of xenobiotics. Polyvinyl chloride (PVC) contains 56% chlorine by weight. Most soils have a low chloride background, and chloride can be determined accurately at very low concentrations (Bergman and Sanik, 1957). Chloride release has been used to detect the biodegradation of various pesticides (Bartha, 1990) and was used by Yabannavar and Bartha (1993) to detect whether weight losses from plasticized PVC formulations were due to PVC or to plasticizer degradation. As only marginal chloride releases from PVC were detected in these experiments, chloroacetic acid at one tenth of the PVC plastic concentration was used as a positive control for the monitoring procedure. From 1 mg chloroacetic acid per g soil, 0.36 mg was degraded after one week, as determined by chloride release. The positive control indicated that the monitoring technique was suitable to quantify the degradation of 0.5% or more of the PVC plastic applied to soil at 10 mg/g. In this case, however, apparently less than 0.1% w/w of the added PVC was degraded.

Infrared Spectrometry

Infrared (IR) spectrometry of polymer films has been used for identification and quality control purposes (Haslam and Willis, 1965; Sadtler Research Laboratories, 1980). Biodegradation may change the IR spectra of films by decreasing some spectral bands and introducing new ones, as some carbon–carbon bonds are broken and functional groups such as hydroxyl or carbonyl introduced. IR spectrometry was used for studying changes that polyethylene (PE)–starch films underwent when exposed in aquatic environments (Imam et al., 1992). In soil, Goheen and Wool (1991) used IR spectrometry to measure, by decrease of the C–O band, the decrease in the starch content of PE–starch blends (29–67% starch, w/w). Changes in the PE spectrum were also monitored. Upon retrieval from soil, in addition to cleaning, the films had to be processed under heat and pressure to reduce their opacity prior to the spectrometric measurements. With this processing, IR spectrometry proved to be an accurate way to follow the decline of starch in PE–starch films. The bands monitored to detect changes in PE did not change significantly in this experiment.

Gas Chromatography

PVC-type plastics contain plasticizers, a great variety of which are used. Esters of phthalic acid or adipic acid, the most common plasticizers, can be selectively extracted from the formulated PVC polymer and determined by gas chromatography (GC) techniques. Alternately, the whole PVC formulation may be dissolved in a suitable solvent such as methyl ethyl ketone (MEK). The PVC then is precipitated by using cold methanol and quantified by weighing. The plasticizers remain in solution and can be determined by GC (Yabannavar and Bartha, 1993). Commonly the plasticizers, especially the adipate esters, are degraded more rapidly than the PVC resin. Again, GC work can help in interpreting weight loss by measuring loss of plasticizer by an independent technique. Interestingly, although plasticizer loss initially is rapid, it does not go to completion. Physical inaccessibility of some of the plasticizer trapped among the PVC fibers appears to be the likeliest explanation for this phenomenon.

Thermogravimetry

Thermogravimetry measures weight loss by volatilization from a test sample with rising temperature. To prevent oxidation, the sample is heated in an inert atmosphere. In a composite polymer sample, the amount of certain filler ingredients can be determined by this technique. Andrady (1990) used this procedure to determine the loss of starch from PE–starch formulations in the marine environment. Only the proper cleanup of films exposed to the soil environment would be required to apply this technique to biodegradation studies in this environment.

REFERENCES

Albertsson, A. C. 1978. Biodegradation of synthetic polymers. II. A limited microbial conversion of ^{14}C in polyethylene to $^{14}CO_2$ by some soil fungi. *Journal of Applied Polymer Science* 22:3419–33.

Albertsson, A. C. 1980. The shape of the biodegradation curve for low and high density polyethylenes in prolonged series of experiments. *European Polymer Journal* 16:623–30.

Alexander, M. 1994. *Biodegradation and Bioremediation.* New York: Academic Press.

Aminabhavi, T. M., R. H. Balundgi, and P. E. Cassidy. 1990. A review on biodegradable plastics. *Polymer Plastic Technology and Engineering* 29:235–62.

Andrady, A. L. 1990. Weathering of polyethylene (LDPE) and enhanced photodegradable polyethylene in the marine environment. *Journal of Applied Polymer Science* 39:363–70.

Atlas, R. M. and R. Bartha. 1993. *Microbial Ecology: Fundamentals and Applications*, 3rd Ed. Redwood City, CA: Benjamin Cummings.

Bartha, R. 1990. Isolation of microorganisms that metabolize xenobiotic compounds, in *Isolation of Biotechnological Organisms from Nature*, ed. D. P. Labeda. New York: McGraw-Hill

Bartha, R. and D. Pramer. 1965. Features of a flask and method for measuring the persistence and biological effects of pesticides in soil. *Soil Science* 100:68–70.

Bartha, R., I.-S. You, and A. Saxena. 1983. Humus-bound residues of phenylamide herbicides: Their nature, persistence and monitoring, in *Pesticide Chemistry—Human Welfare and the Environment*, Vol. 3, ed. J. Miyamoto and P. C. Kearney. Oxford: Pergamon Press.

Benedict, C. V., W. J. Cook, P. Jarrett, J. A. Cameron, S. J. Huang, and J. P. Bell. 1983a. Fungal degradation of polycaprolactones. *Journal of Applied Polymer Science* 28:327–34.

Benedict, C. V., J. A. Cameron, S. J. Huang, and J. P. Bell. 1983b. Polycaprolactone degradation by mixed and pure cultures of bacteria and yeast. *Journal of Applied Polymer Science* 28:335–42.

Bergman, J. C. and J. Sanik, Jr. 1957. Determination of trace amounts of chlorine in naphtha. *Analytical Chemistry* 29:241–43.

Booth, G. H. and J. A. Robb. 1968. Bacterial degradation of plasticized PVC—effect on some physical properties. *Journal of Applied Chemistry* 18:194–97.

Goheen, S. M. and R. P. Wool. 1991. Degradation of polyethylene–starch blends in soil. *Journal of Applied Polymer Science* 42:2691–701.

Haslam, J. and H. A. Willis. 1965. *Identification and Analysis of Plastics.* Princeton, NJ: D. Van Nostrand.

Imam, S. H., J. M. Gould, S. H. Gordon, M. P. Kinney, A. M. Ramsey, and T. R. Tosteson. 1992. Fate of starch-containing plastic films exposed in aquatic habitat. *Current Microbiology* 25:1–8.

Johnson, K. E., A. L. Pometto III, and Z. L. Nikolov. 1993. Degradation of degradable starch–polyethylene plastics in a compost environment. *Applied and Environmental Microbiology* 59:1155–161.

Klausmeier, R. E. 1972. Results of the second interlaboratory experiment on biodeterioration of plastics. *International Biodegradation Bulletin* 8:3–7.

Kuhn, E. P. and J. M. Suflita. 1989. Dehalogenation of pesticides by anaerobic microorganisms in soil and groundwater — a review, in *Reactions and Movement of Organic Chemicals in Soils*. Special Publication no. 22, pp. 111–80. Soil Science Society of America.

Lee, B., A. L. Pometto, A. Fratzke, and T. Bailey. 1991. Biodegradation of degradable plastic polyethylene by *Phanerochaete and Streptomyces species*. *Applied and Environmental Microbiology* 57:678–85.

Lefebre, F., C. Davaid, and C. Vander Wauven. 1994. Biodegradation of polycaprolactone by microorganisms from an industrial compost of household refuse. *Polymer Degradation and Stabilization* 45:347–53.

Mergaert, J., A. Webb, C. Anderson, A. Wouters, and J. Swings. 1993. Microbial degradation of poly(3-hydroxybutyrate) and poly(3-hydroxybutyrate-Co-3-hydroxyvalerate) in soils. *Applied and Environmental Microbiology* 59:3233–38.

Pramer, D. and R. Bartha. 1972. Preparation and processing of soil samples for biodegradation studies. *Environmental Letters* 2:217–24.

Sadtler Research Laboratories. 1980. *The Infrared Spectra Atlas of Monomers and Polymers*. Philadelphia, PA: Sadtler Laboratories (Division of Biorad).

Sharabi, N. E.-L. and R. Bartha. 1993. Testing of some assumptions about biodegradability in soil as measured by carbon dioxide evolution. *Applied and Environmental Microbiology* 59:1201–5.

Shen, J. 1994. Determination of biodegradability in soil by measuring net CO_2 evolution in biometer flasks. *Ph.D. dissertation*. Rutgers University, New Brunswick, NJ.

Sielicki, M., D. D. Focht and J. P. Martin. 1978. Microbial degradation of ^{14}C polystyrene and 1,3-diphenylbutane. *Canadian Journal of Microbiology* 24:798–803.

Wendt, T. M., A. M. Kaplan, and M. Greenberg. 1970. Weight loss as a method for estimating the microbial deterioration of PVC film in soil burial. *International Biodegradation Bulletin* 6:139–43.

Yabannavar, A. and R. Bartha. 1993. Biodegradability of some food packaging materials. *Soil Biology and Biochemistry* 25:1469–75.

Yabannavar, A. and R. Bartha. 1994. Methods for assessment of biodegradability of plastic films in soil. *Applied and Environmental Microbiology* 60:2717–722

Chapter *5*

Biodegradation Testing of Polymers in Compost

■ MICHAEL A. COLE
  University of Illinois

INTRODUCTION

Composting is a waste management process in which biologically degradable organic wastes are converted into relatively stable organic end products and carbon dioxide (CO_2). As a result of successful stabilization by composting, organic waste can be used as a soil amendment product to stimulate plant growth. Both household and commercial-scale composting are becoming more attractive as solid waste disposal options, given diminishing landfill capacity, increasing legislative pressure on incineration, and improving composting technology.

The chemical, biological, and physical conditions of the composting process are substantially different from those encountered in aquatic and soil environments. These differences are important because compost can affect the fate of degradable polymers in ways that would not be anticipated from test results obtained in aquatic and soil media. The purpose of this chapter is to describe the significance of degradable plastic polymers in compost, to outline test systems by which degradability of plastics in compost can be evaluated, and to present some examples of the performance of degradable plastics during composting.

SIGNIFICANCE OF PLASTICS IN COMPOST

Sources of compostable material include food, paper, and yard waste as well as feedstock from municipal solid waste (MSW) facilities. As interest in minimizing landfill disposal of biodegradable organic wastes has increased during the past decade, it has become apparent that composting of such materials is an environmentally sound and a potentially economically viable method of converting the wastes into a useful product. However, convenient handling of such materials by homeowners and solid waste collectors is relatively difficult if the materials are not placed in waterproof bags; odors and liquid spills are frequent complaints. When nondegradable plastic bags are used, compost producers may have difficulty selling the finished compost because the presence of the plastic makes the compost unacceptable to users.

Regulations in some European countries and in several states in the United States that limit the amount of inert materials (including plastics) in end-product compost. To resolve this problem, biodegradable plastic compost bags were suggested as an alternative to nondegradable bags. Ideally, the degradable compost bags would not interfere with compost production and would not be present as visible plastic in finished compost. Further, composting of restaurant wastes is not practical unless all the waste, including all plastic food wrappings, beverage containers, and eating utensils, is biodegradable. Restaurant customers generally will not separate degradable and nondegradable components of the waste stream.

Mineralization is the conversion of an element (e.g., carbon) from an organic form to an inorganic state as a result of microbial activity. The ideal end result of full degradation of plastic-containing waste by composting is destruction of the plastic, depolymerization to monomers, and conversion of the monomers to CO_2 and microbial biomass, thereby leaving no plastic residues and no concerns for environmental or human health effects. A solution to handling and separation problems is to use biodegradable plastics that could be mingled with the food wastes and would disappear during the composting process. However, in addition to producing materials whose performance is satisfactory, one must also consider the possible consequences to the environment of using such materials. If a plastic is not mineralized completely by microbial or chemical processes, then a residue will remain. General concerns about residues in the environment have been raised, such as the Office of Technology Assessment claim that "Many important questions about the rate and timing of degradation in different environments and about the environmental safety of degradation products have either not been addressed or the research is only now underway" (OTA, 1989).

There are a number of quite different perspectives among concerned groups and individuals on the extent of plastics degradation that is satisfactory or noteworthy. The issue of what is a satisfactory endpoint of degradation is not simply a matter for esoteric debate. A clear consensus on this point is vital because such a decision, particularly if codified by national legislation, will set the direction for further research on these materials. If the ultimate goal of degradable plastic technology is to make plastics disintegrate into inconspicuous particles, irrespective of chemical identity and quantity of residues, then very substantial progress toward this goal has been made already. If the goal is to make plastics that completely biodegrade, then substantial improvements in degradation technology as well as plastic composition, performance, and costs are still needed.

COMPOSTING SYSTEMS

Characteristics of Composts

Compost mixes typically include of a wide variety of naturally occurring polymers. However, the production of enzymes that degrade natural polymers does not always occur uniformly throughout the compost processing cycle. For example, cellulase, β-glucoside, and esterase activities were much higher in a mature MSW compost than they were earlier in the process (Herrmann and Shann, 1993). As many degradable plastics have hydrolyzable linkages such as esters or amides, these polymers may degrade better in mature compost than during other phases.

Maximum soil and water temperatures in temperate climates range between 5°C and 35°C. In contrast, commercial composting systems typically generate temperatures of 55°C or higher as an integral part of composting, as elevated temperatures are an operational requirement to achieve destruction of microbial pathogens, parasites, adult insects, their larvae and eggs, and weed seeds. A typical thermal profile in compost is a temperature increase from ambient to 55°C to 65°C within 2 to 7 days, followed by a thermophilic phase ($\geqslant 55$°C) for several days to several weeks. The thermophilic phase is followed by a maturation period of several weeks with temperatures around 35°C to 45°C, ending with a cooling period to ambient that lasts from a few weeks to several months. For plastics that are susceptible to chemical degradation, higher temperatures in compost may increase their degradation rate when compared to their degradation rate in lower-temperature environments such as soil or water.

Compost has a relatively narrow pH range, with initial values of 6 to 7 and values of 8 to 8.5 after the thermophilic phase. If the oxygen (O_2) supply is inadequate, pH values of 4.5 to 5.5 are found. Depending on the starting material, the soluble salt content of compost is much higher than that of soil, and the concentration of specific soluble ions such as ammonium, calcium, magnesium, and potassium is elevated when compared to soil.

Microbial populations and activity in compost are higher than in soil and water (except for highly eutrophic lakes), and the microbial species distribution is substantially different from that in soil or water. Because of changes in temperature and substrate availability during the composting process, microbial succession occurs rapidly, with very different organisms being abundant (and probably active) during the different phases of the process.

Compost is a more complex chemical environment than soil and water because of the relatively high concentrations of low molecular weight organic and inorganic solutes in compost that could enhance or inhibit chemical degradation. Further, Cole (1990a, b) has suggested that:

- Leaching of degradation catalysts into compost from a degradable starch–polyethylene blend can cause poor degradability in compost.
- Accumulation of microbial biofilm on plastic surfaces may reduce degradation potential because oxygen (required for production of free radicals that carry out chain scission) would be consumed by the biofilm.

Biofilm accumulation on thermally degraded PE was found to inhibit chemical degradation during compost incubation (Johnson et al., 1993; Pometto et al., 1995). These results suggest that microbial colonization of plastics will slow chemical reactions of plastics in compost.

Large-Scale Municipal Compost Systems

There are numerous designs for large-scale composting systems for municipal solid waste. Specific processing systems and vendors have been discussed in detail elsewhere (e.g., Haug, 1993; Hoitink and Keener, 1993). A typical processing sequence for MSW compost is shown in Figure 5.1. Incoming materials are sorted, and recyclables are removed. The residual material is continuously fed into a drum, in which the residence time is 3 days. The material is mechanically screened (15 cm cutoff), and oversize material is either landfilled or used as incinerator fuel. This step removes large pieces of plastic, metal, plastic-coated cardboard, and wood. Material passing through the screen is transferred into concrete trenches with bottom

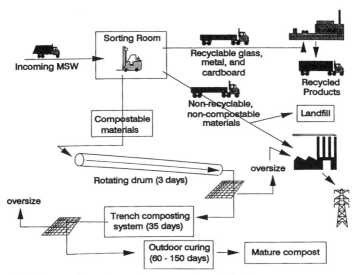

Figure 5.1. Processing scheme for unsorted municipal solid waste (facility located at St. Cloud, MN).

aeration and moisture control, and is turned daily. The turning process advances the material in the trench, from which it exits in about 35 days. A second mechanical screening (3 cm cutoff) is performed, and material passing the screen is placed in outdoor windrows for curing. Oversize material is discarded. The stabilized end-product compost from this plant is used as landfill cover; however, the finished compost from other composting facilities is sold as a soil amendment for horticultural, agricultural, or landscape architecture (e.g., home and commercial) use.

There are variations in mechanical design and processing schemes among various vendors of composting equipment, but both batch process and continuous feed systems have several characteristics:

- Oxygen is supplied by forced air or vacuum, with the result that air-filled pores have a relatively high oxygen content. However, anaerobic conditions exist in the interior of particles where diffusion is the primary mechanism for oxygen transport. Because the composting mass has both aerobic and anaerobic regions, microbial biodiversity is high, and a wide range of chemical environments exists with respect to oxygen content and redox potential.
- Some method of continuous or periodic turning is employed to ensure thorough mixing and to achieve particle size reduction. As a result of sheer forces generated during mixing and the stress imposed on

plastics as the composting mass settles, the plastics are subjected to much more mechanical stress than they would encounter in soil or in water (with the exception of beached materials that are subjected to wave action).

- Water is added as necessary to maintain a moisture content of about 50% (w/w) to promote and maintain rapid microbial degradation. The more controlled moisture regime during composting is a great contrast to the situation of soils, as the moisture content of field soils fluctuates on a daily and a seasonal basis.

Typical depths of composting material at commercial composting operations are 1.5 to 2.5 m, in contrast to surface soils, whose depth can be one meter or less. The greater depth of compost than of soil places more stress on the buried plastic, with the result that mechanical degradation in compost will be greater than that encountered in field soil or laboratory-scale compost reactors.

EVALUATION OF BIODEGRADATION IN COMPOST

Polymer Chemistry

As indicated in Chapter 15, overall biodegradation potential varies according to the vulnerability of chemical bonds within each type of polymer. For example, thermal degradation of polyolefins is very slow at temperatures less than 50°C. Therefore, these materials would not be expected to degrade rapidly if buried under most natural conditions in soil. Thermosensitive polymers are potentially degradable during the thermophilic phase ($\geqslant 55°C$) of composting.

Many biodegradable plastic formulations are based on ester-linked monomers (Helmus, 1990). This group includes polyhydroxybutyrate-valerate (PHBV) and polylactic acid, among others. These polymers are biodegradable because the ester linkages of such polyesters are susceptible to both enzymatic attack and chemical hydrolysis. Composting with insufficient oxygen available will generate slightly acidic conditions (pH $\sim$ 5), whereas completely aerobic systems can become slightly alkaline (pH 8.0–8.5). In conjuction with elevated temperatures, rapid chemical hydrolysis of ester linkages can occur.

Test Systems and Measurements

Degradation of plastics can be assessed in laboratory or pilot-scale reactors that attempt to simulate large-scale composting systems. Bench-scale laboratory reactors typically contain only a few liters of composting material, whereas pilot-scale systems contain a cubic meter or more. Each test system has advantages and disadvantages, and the choice of which to use is decided largely by the exact question that is being considered. Laboratory and pilot-scale systems are used primarily for preliminary screening of potential materials, for evaluating the effects of specific system parameters such as moisture content or temperature on degradation, and for research studies where quantitative recovery of samples is desired.

A variety of measurements are commonly used to assess degradation of plastics. These include changes in molecular weight distribution or mechanical properties, loss of weight during exposure, disappearance, O_2 consumption, or CO_2 release. In fully biodegradable plastics, the following sequence is seen (Figure 5-2):

- Loss of mechanical strength due to hydrolysis of a relatively small fraction of the susceptible bonds.
- Visible erosion of the plastic surface and weight loss.
- Disintegration of the initial test specimen into progressively smaller and eventually invisible fragments.
- Microbial degradation of the relatively low molecular weight oligomers and monomers of the plastic.

In other cases, the degradation process ceases long before suitable substrates for complete microbial degradation are formed, leaving fragments

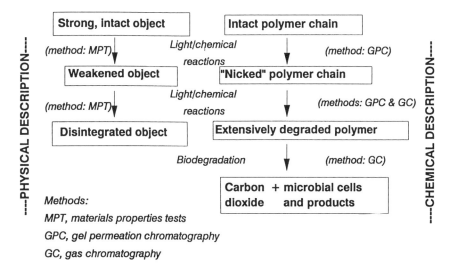

Figure 5.2. Stages in polymer degradation and associated measurements.

of plastic. Further, the disappearance of samples cannot be used as evidence of biodegradability in field studies. Test materials may "disappear" in natural systems purely as a result of physical processes or animal activity (Wiegert and Evans, 1964). If inconspicuous nondegradable fragments or residues are formed in compost, then it is potentially necessary to conduct toxicological studies to ensure that the residues are not harmful. Of course, the simplest approach from a toxicological and environmental standpoint is to develop plastics that leave no residues that may elicit concern.

Many designs for laboratory-scale compost reactors have been proposed (e.g., Ashbolt and Line, 1982; ASTM, 1994a, b; Frankos et al., 1982; Griffin, 1976; Haug and Ellsworth, 1991; Hogan et al., 1989; Jeris and Regan, 1973). All of these systems have common features, and a typical example is presented in Figure 5.3. In most cases, the objective of work in composting studies is to determine some form of mass balance (e.g., by evolution of carbon from the plastic as CO_2 or by measuring weight loss of material over time).

There are substantial differences in the sophistication of the process control and outlet monitoring systems, as well as the blend of materials used for the composting matrix. Practically speaking, simpler designs are less expensive and allow the researcher to construct enough units to perform side-by-side comparisons of materials with adequate replication to identify significant differences among materials. The principal features of most composting reactors are:

- Incoming air is passed through an alkaline trap (usually a solution of barium, sodium, or potassium hydroxide) to remove CO_2 from the air by converting it to nonvolatile bicarbonate ion or precipitating it as barium carbonate. The scrubber also serves to humidify the air before it enters the compost.
- Exit air is analyzed for CO_2 or O_2 on an intermittent or continuous basis by capturing CO_2 in alkali, by infrared gas determination of CO_2 and O_2, or by various other methods.
- The materials in the reactor are mixed periodically to maintain relatively constant moisture throughout the vessel contents. Maintaining even moisture distribution is often difficult. For example, if the inlet air is dry, the compost on the inlet side of vessel tends to dry out. In contrast, if the inlet air is humidified, compost near the air inlet may become oversaturated with water.
- Heat production is usually less than heat loss in small-vessel systems, with the result that relatively small vessels will not maintain thermophilic conditions. In such cases, incubation is carried out in a temperature-programmed chamber or water bath, or the vessel is enclosed within a heating mantle.

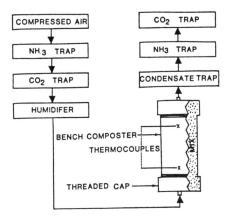

Figure 5.3. Schematic diagram of bench-top composting unit (Frankos et al., 1982).

As noted above, a wide variety of compostable materials have been used as feedstock for the compost reactors. Raw MSW is rarely used. MSW is highly variable in composition on a day-to-day and a seasonal basis and does not provide a reproducible feedstock. Its particle size is often too large for small-scale reactors. MSW also is a biohazard, potentially containing a wide variety of pathogenic bacteria, fungi, and viruses. Therefore, the use of raw MSW is not recommended.

PERFORMANCE OF PLASTICS

General Concepts

General concerns in assessing plastic biodegradation are well summarized by the following questions (Gu et al., 1993a):

- What time period is required for plastic decomposition and mineralization in relevant disposal environments such as compost, wastewater treatment facilities, and landfills?
- Does the loss of properties and weight of a plastic article result from a biological process (biodestructability), with the subsequent conversion of carbon (at least in part) to biological products such as biogas, or is the decomposition nonbiologically mediated, producing recalcitrant by-products?
- Does the decomposition process result in the formation of toxic metabolic intermediates?

- Does the decomposition process occur with complete mineralization of the disposable plastic article?
- What effects will variable environmental exposure conditions between different waste treatment facilities (e.g., between different compost facilities) have on the measured rate of plastic decomposition?

Further, three major requirements must be met before full degradation of most synthetic polymers will occur:

- The polymer must be quantitatively converted to lower molecular weight products by nonbiological reactions because the common packaging polymers are relatively immune to enzymatic attack. Existing technologies for this purpose include incorporation of photosensitive, chemically labile, or enzymatically susceptible sites into the polymer chain or the use of various additives to accelerate chemical scission of the polymer (e.g., Guillet, 1987; Harlan and Nicholas, 1987; Maddever and Campbell, 1990).
- A high percentage of the end products of these reactions must fall within the size domain of molecules that are susceptible to microbial processing.
- The environment in which the polymer is placed cannot inactivate the abiotic degradation system and must support the activity of microbes that metabolize the plastic degradation products.

Plastics in compost should meet definite performance standards. These products must have sufficiently degraded so that they are no longer visible as fragments by completion of the composting process. This requirement is established by the low tolerance of compost users for conspicuous foreign materials such as plastics in the finished product. Further, contaminant criteria used in some states can preclude the use of compost containing inorganic or organic pollutants.

The time available for plastic degradation can be established if the stabilization rate of the compost is known. First-order decay constants (k) ranged from 0.069 kg/kg-day for straw to 0.190 kg/kg-day for a mixture of brush, grass, and leaves when processed in a laboratory compost unit (Keener et al., 1993). For appropriate compost mixtures, these values can be used to calculate the time required to achieve stabilization of the compost (i.e., typically when the initial organic content has been reduced by 50%). When compost stabilization data are combined with plastic degradation data, a useful plastic degradation and compost stabilization profile such as that of Figure 5.4 can be obtained. In reality, however, compost degradation of plastics often is not sufficient to remove all plastic fragments from the compost. For example, starch-containing yard waste collection bags had

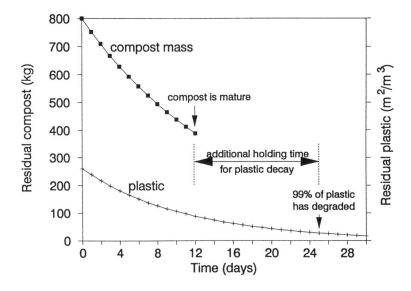

Figure 5.4.   Illustrative profile of compost stabilization and plastic degradation.

lost only 40% of their initial strength during a 64-week exposure in compost; the estimated residence time of the plastic is two to three years (Leonas et al., 1994). The compost was mature by 48 weeks, and substantial amounts of visible plastic were still present. In general, if the decay rate of the plastic is less than the decay rate of the composting material, an increase can occur with time in the percentage of plastic in the compost.

Performance of Major Plastics

The current use of billions of pounds of plastics such as polyethylene (PE), polypropylene (PP), polystyrene (PS), and polyvinylchloride (PVC) is a testimonial to the excellent overall performance of these materials. Few new materials have all the desirable barrier properties, cost-effectiveness, or processability of these plastics, nor have many of the newer materials been adequately tested or certified for food-contact or long-term food storage. When considering the relative merits of PE, PP, PS, or PVC as starting points for fully degradable plastics, one should note that these plastics are partially degradable when chemical stabilizers are omitted, and/or when catalysts are used to promote photodegradation. However, only limited biodegradation of PS and PE occurred when the polymers were pre-degraded by light or chemical reactions (Klemchuk, 1989); the extent of biodegradation was much lower for PS compared to PE. Most published work indicates that PS oligomers that are larger than trimers are not completely biodegradable, and that only limited biotransformation of larger oligomers takes place (Haraguchi and Hatakeyama, 1980).

Some investigators have suggested that PE, PP, and PS are "biodegradable" on the basis, for example, of the conversion of a small percentage of material to CO_2. Based on the formation of small amounts of characteristic products of microbial metabolism in short-term studies, it might be assumed that all of the plastic eventually could be mineralized. Inspection of the data indicates that this assumption is not correct.

Although early results suggested that some microbial transformation of plastics had taken place, the observed degradation kinetics are most simply interpreted as rapid metabolism of a small fraction of the total material followed by substrate depletion (i.e., a point at which little or no further degradation occurs). The recent literature indicates that only a small fraction of extensively photodegraded PE can be mineralized. Chapter 9 describes case studies of the fate of PE in compost. An exception to this statement is the work of Albertsson et al. (1993), who found that thermally degraded low density polyethylene (LDPE)–starch blends evolved 2.5% of total C as ^{14}C-labeled CO_2 when incubated with fungal cultures in a liquid medium for 600 days. In the same study, only 0.7% of ^{14}C was evolved as $^{14}CO_2$ from thermally degraded LDPE without corn starch. In neither case did the yield of CO_2 approach the amount expected from full biodegradation of the plastic. These authors subjected the LDPE to a temperature of 100°C for 6 days prior to exposure to fungi. It is difficult to imagine an actual land disposal situation where such heat treatment could occur; therefore, this work also supports the contention that PE is poorly biodegradable in compost.

Microbial growth on PE is probably due to selective utilization of the very low molecular weight fraction and/or biodegradable additives in PE films. For example, Weiland found that 28% to 64% of the carbon from PE that was subjected to thermal degradation and then added to a suspension of compost-derived organisms was released as CO_2 (Weiland et al., 1995).

METHODOLOGICAL ISSUES

Background Carbon Dioxide

Degradation of the polymer (e.g., plastic) in compost is typically determined by measuring CO_2 produced or oxygen consumed in excess of the quantities produced or used in the compost without plastic. Conducting these experiments with compost is substantially more difficult than similar experiments with soil or aquatic systems because the biological activity and quantities of CO_2 produced and O_2 consumed are manyfold higher in compost than in soil or water, as shown in the examples in Table 5.1. A similar difference in scale would be expected for CO_2 evolution, because O_2 uptake and CO_2 evolution are coupled processes in aerobic organisms. As a result of the high metabolic activity in the compost, plastic degradation can be assessed only if the plastic decays very rapidly or if it degrades during the later stages of composting when evolution of CO_2 from the compost is relatively low. In order to detect mineralization of the plastic, the degradation rate (measured as CO_2 evolution) of the plastic has to be high enough to result in a significant increase in CO_2 evolution over the nonamended compost. If the plastic degrades relatively slowly, CO_2 evolution from the plastic cannot be detected as a significant increase over plastic-free compost. Thus, the plastic may actually be mineralized, but evidence of the degradation cannot be obtained. To illustrate the difficulty of working with compost compared to soil, consider the following examples:

- A soil–plastic mix containing 90% soil and 10% plastic (w/w), the soil having a respiration rate (as measured by CO_2 evolution) of 0.035 mg CO_2 g^{-1} soil hr^{-1} (equivalent to an O_2 uptake rate of 0.048 mg O_2 g^{-1} soil hr^{-1}, assuming a 1:1 molar ratio between CO_2 production and O_2 consumption).
- A compost–plastic mix containing 90% compost and 10% plastic (w/w), the compost having a respiration rate of 4.0 mg CO_2 g^{-1} hr^{-1} (equivalent to 5.5 mg O_2 consumed g^{-1} h^{-1}).

Table 5.1. Oxygen consumption rates by mineral soil, forest litter, and composts

Mineral	O_2 consumption (mg O_2g^{-1} dry wt. hr^{-1})	Reference
Soil	0.02	Bunt and Rovira, 1955
Forest litter, 2 weeks after deposition, 10°C	0.18–0.30	Howard, 1967
Forest litter, 53 weeks after deposition	0.04–0.11	Howard, 1967
Immature MSW compost, 50°C	2.5–5.5	Regan and Jeris, 1970
Immature MSW compost, 37°C	1.4	Iannotti et al., 1993
Mature MSW compost, 37°C	0.27	Iannotti et al., 1993

If the plastic must degrade at a rate such that total CO_2 production (soil or compost, plus plastic) has to be 40% greater than CO_2 evolution from the soil or compost alone (i.e., 40% being the minimum increase in CO_2 detectable when compared to the soil or the compost alone), then the necessary rate of plastic degradation can be calculated from the formula:

$$(\text{Soil rate} \times \text{Mass fraction of soil}) + (\text{Plastic rate} \times \text{Mass fraction of plastic}) = 140\% \times \text{Soil rate}$$

The solution for this equation indicates that the rate of CO_2 evolution from the plastic must be five times greater than CO_2 evolution from the soil alone. Based on a biodegradable plastic containing 65% carbon (w/w) and the carbon being partitioned into 80% C evolved as CO_2 and 20% retained as microbial biomass, all the plastic would be degraded in slightly over one year. Even if this plastic degrades quite slowly in soil, its degradation can be accurately measured. However, the same plastic would have to be degraded completely in an immature compost (e.g., $5.5\,\text{mg O}_2\,\text{g}^{-1}\,\text{hr}^{-1}$) in about 12 hr. This is an extremely high rate that would not be easily attained. Therefore, it is possible to measure degradation of only the most rapidly decaying plastics in an immature compost. If a mature compost were used as the test medium (e.g., $0.27\,\text{mg O}_2\,\text{g}^{-1}\,\text{hr}^{-1}$), the time allowed for full degradation would be about 178 hr, which is an attainable performance. For example, Gu measured degradation of cellulose and cellulose acetate in mature compost (Gu et al., 1993b), finding that the residence time for the cellulose, based on CO_2 evolution, was about 8 days, and the residence time of cellulose acetate, based on visual disappearance of test samples, was about 8 to 12 days. Issues with using mature compost are outlined below.

Use of Mature Compost

A difficulty with using mature compost as a test matrix is that the active microbes and the chemical and thermal conditions of mature compost are much different from those found during the high-activity thermophilic phase of composting. For example, cellulase, β-glucoside, and esterase activities were much higher in a mature MSW compost than earlier in the process (Herrmann and Shann, 1993). Therefore, using mature compost may not accurately reflect the biological or chemical conditions that plastics would encounter during the early stages of the composting process. If the conditions are substantially different, then the laboratory-based results may not transfer well to real-world composting situations.

Partial Oxidation and Organic Coupling

In soil systems, extensive mineralization of a wide range of simple aliphatic and aromatic hydrocarbons and their hydroxyl and carboxyl derivatives has been demonstrated many times (Atlas, 1984). These compounds are likely partial degradation products of most plastics. Therefore, upon extensive biodegradation of a material in soil, mineralization to CO_2 can be observed. In compost, however, a substantially different situation may occur. Many simple aromatic compounds are not mineralized in compost, but instead are partially oxidized and then covalently coupled to stable organic constituents such as humic and fulvic acids (Benoit and Barriuso, 1995). As a result of covalent coupling, the degradation of polymers may not always be accurately assessed by mineralization (e.g., release of CO_2) alone.

FUTURE OF COMPOSTING PLASTICS

Plastics and other polymer products that are found to be truly biodegradable in compost have a potentially large market because of widespread public appeal and their role in reducing the use of incineration and landfills. These products can be evaluated in the context of recommendations from a task force on environmental advertising (State of Minnesota, 1991). As indicated by the task force, "it may be appropriate to make claims about the biodegradability of a product when that product is disposed of in a waste management facility that is designed to take advantage of biodegradability (such as a municipal solid waste composting facility) *and* the product at issue will *safely* break down at a sufficiently rapid rate and with enough completeness when disposed of in that system to meet the standards set by any existent state or federal regulation." Therefore, to help ensure the market for compostable polymers, manufacturers should work not only on evaluating product toxicity and performance but also on development of compost disposal systems that facilitate polymer biodegradation. Case studies of composting plastics are provided in Chapter 9.

REFERENCES

Albertsson, A. C., C. Barenstedt, and S. Karlsson. 1993. Increased biodegradation of a low-density polyethylene (LDPE) matrix in starch-filled LDPE materials. *Journal of Environmental Polymer Degradation* 1:241– 45.

Ashbolt, N. J. and M. A. Line. 1982. A bench-scale system to study the composting of organic wastes. *Journal of Environmental Quality* 11:405– 8.

ASTM. 1994a. D5509: Practice for exposing plastics in simulated compost environment, in *Annual Book of ASTM Standards*, Vol. 8.03. Philadelphia, PA: American Society for Testing and Materials.

ASTM. 1994b. D5512: Practice for exposing plastics to a simulated compost environment using an externally heated reactor, in *Annual Book of ASTM Standards*, Vol. 8.03. Philadelphia, PA: American Society for Testing and Materials.

Atlas, R. M. 1984. *Petroleum Microbiology*. New York: Macmillan.

Benoit, P. and E. Barriuso. 1995. Effect of straw composting on the degradation and stabilization of chlorophenols in soil. *Compost Science and Utilization* 3:31–37.

Bunt, J. S. and A. D. Rovira. 1955. The effect of temperature and heat treatment on soil metabolism. *Journal of Soil Science* 6:129–36.

Cole, M. A. 1990a. Constraints on decay of polysaccharide-plastic blends. In *Agricultural and Synthetic Polymers: Utilization and Biodegradability*, ed. J. E. Glass and G. Swift. Washington, DC: American Chemical Society.

Cole, M. A. 1990b. Making polyolefins biodegradable: A systems model. *PMSE Polymer Preprints* 63:872–76.

Frankos, N. H., F. Gouin, and L. J. Sikora. 1982. Using woodchips of specific species in composting. *Biocycle* (May–June):38–40.

Griffin, G. J. 1976. Degradation of polyethylene in compost burial. *Journal of Polymer Science* 57:281–86.

Gu, J.-D., D. Eberiel, S. P. McCarthy, and R. A. Gross. 1993a. A respirometric method to measure mineralization of polymeric materials in a matured compost environment. *Journal of Environmental Polymer Degradation* 1:293–300.

Gu, J.-D., D. Eberiel, S. P. McCarthy, and R. A. Gross. 1993b. Degradation and mineralization of cellulose acetate in simulated thermophilic compost environments. *Journal of Environmental Polymer Degradation* 1:281–91.

Guillet, J. E. 1987. Vinyl ketone photodegradable plastics, in *Degradable Plastics*. Washington, DC: Society of the Plastics Industry.

Haraguchi, T. and H. Hatakeyama. 1980. *Lignin Biodegradation,* Vol. II., ed. T. K. Kirk, T. Higuchi and H. Chang. Boca Raton, FL: CRC Press.

Harlan, G. M. and A. Nicholas. 1987. Degradable ethylene–carbon monoxide copolymer, in *Degradable Plastics.* Washington, DC: Society of the Plastics Industry.

Haug, R. T. 1993. *Compost Engineering.* Boca Raton, FL: Lewis Publishers.

Haug, R. T. and W. F. Ellsworth. 1991. Measuring compost substrate degradability. *Biocycle* (Jan.):56–62.

Helmus, M. N. 1990. The outlook for environmentally degradable plastics, in *Degradable Materials: Perspectives, Issues and Opportunities,* ed. S. A. Barenberg, J. L. Brash, R. Narayan, and A. E. Redpath. Boca Raton, FL: CRC Press.

Herrmann, R. F. and J. R. Shann. 1993. Enzyme activities as indicators of municipal solid waste compost maturity. *Compost Science and Utilization* 1:54–63.

Hogan, J. A., F. C. Miller, and M. S. Finstein. 1989. Physical modeling of the composting ecosystem. *Applied Environmental Microbiology* 55:1082–92.

Hoitink, H. A. J. and H. M. Keener (eds.). 1993. *Science and Engineering of Compost.* Worthington, OH: Renaissance Publications.

Howard, P. J. A. 1967. A method for studying the respiration and decomposition of litter, in *Progress in Soil Biology,* ed. O. Graff and J. E. Satchell. Amsterdam: North-Holland Publishing.

Iannotti, D. A., T. Pang, B. L. Toth, D. L. Elwell, H. M. Keener, and H. A. J. Hoitink. 1993. A quantitative respirometric method for monitoring compost stability. *Compost Science and Utilization* 1:52–65.

Jeris, J. S. and R. W. Regan. 1973. Controlling environmental parameters for optimum composting. *Compost Science* (Jan.–Feb.):10-15.

Johnson, K. E., A. L. Pometto, III, and Z. L. Nikolov. 1993. Degradation of degradable starch–polyethylene plastics in a compost environment. *Applied and Environmental Microbiology* 59:1155–61.

Keener, H. M., C. Marugg, R. C. Hansen, and H. A. J. Hoitink. 1993. Optimizing the efficiency of the composting process, in *Science and Engineering of Composting,* ed. H. A. J. Hoitink and H. M. Keener. Worthington, OH: Renaissance Publications.

Klemchuk, P. P. 1989. Chemistry of plastics casts a negative vote. *Modern Plastics* (Aug.):48–53.

Leonas, K. K., M. A. Cole, and X.-Y. Xiao. 1994. Enhanced degradable yard waste collection bag behavior in a field-scale composting environment. *Journal of Environmental Polymer Degradation* 2:253–61.

Maddever, W. J. and P. D. Campbell. 1990. Modified starch based environmentally degradable plastics. In *Degradable Materials: Perspectives,*

Issues and Opportunities, ed. S. A. Barenberg, J. L. Brash, R. Narayan, and A. E. Redpath. Boca Raton. FL: CRC Press.

OTA. 1989. *Facing America's Trash: What Next for Municipal Solid Waste.* U.S. Congress, Office of Technology Assessment, Document OTA-O-424. Washington, DC: U.S. Government Printing Office.

Pometto, A. L., III, K. E. Johnson, and M. Kim. 1993. Pure-culture and enzymatic assay for starch–polyethylene degradable plastic biodegradation with *Streptomyces species. Journal of Environmental Polymer Degradation* 1:213–21.

Regan, R. W. and J. S. Jeris. 1970. A review of the decomposition of cellulose and refuse. *Compost Science* (Jan.–Feb.):17–20; references in (July–Aug.):32.

State of Minnesota. 1991. *The Green Report II—Recommendations for Responsible Environmental Advertising.* St. Paul, MN: Office of the Attorney General.

Weiland, M., A. Daro, and C. David. 1995. Biodegradation of thermally oxidized polyethylene. *Polymer Degradation and Stabilization* 48:275–89.

Wiegert, R. G. and F. C. Evans. 1964. Primary production and the disappearance of dead vegetation on an old field in southeastern Michigan. *Ecology* 45:49–63.

PART 2 | *Ecological Toxicity and Fate Assessment*

Chapter 6

Polycarboxylates and Polyacrylate Superabsorbents

Part I: Polycarboxylates

■ JOHN D. HAMILTON,[a] ISADORE J. MORICI,[b] AND MICHAEL B. FREEMAN[b]
S. C. Johnson Wax[a] and Rohm & Haas Company[b]

INTRODUCTION

Under regulatory efforts to minimize eutrophication of natural waters, the use of phosphate-containing laundry detergents has been restricted or banned. When used to replace phosphate dispersants in laundry detergents, polycarboxylates prevent redeposition and encrustation of inorganic salts on fabrics during cleaning. Polycarboxylates can also prevent inorganic encrustation from water in cooling systems, and will function as process aids to enhance product flow during processing. Other applications of polycarboxylates include their use in pigments, drilling muds, paints, textile manufacturing, and leather processing.

As part of the broad class of acrylic polymers, polycarboxylates are characterized by their relatively low molecular weight (i.e., average MW less than 100,000) and carboxylate functional groups. They are highly soluble in water and are typically supplied as 40% to 50% solutions (by weight); dry powders are also available. Commonly used polycarboxylates are a sodium salt acrylic acid homopolymer with an average MW of 4,500 and a sodium salt acrylic acid–maleic acid copolymer with an average MW of 70,000. Other polycarboxylates with similar physical-chemical properties include copolymers and terpolymers of acrylic acid with methacrylic acid, sulfonic acids, hydroxyacrylates, and/or acrylamides.

Recent trends in the detergent market have led to substantial use of polycarboxylate dispersants in home laundry detergent and related cleaning formulations. For example, the total end use of polycarboxylates in the United States recently was estimated to be 75 million kilograms (on a polymer basis), with 26.3 million kilograms used in detergents and cleaners (SRI, 1992). Household laundry detergents are the major end use for polycarboxylates in the United States. The percent by weight of polycarboxylates in these products ranges worldwide from approximately 0.5% to 5% by weight, depending on the formulation and the intended function of the polymer.

Given the relatively high commercial use levels of polycarboxylates, it is important to consider their ecotoxicity and ecological fate. A substantial amount of important information is available for evaluation of the ecotoxicity and the ecological fate of polycarboxylates.

ECOLOGICAL FATE

Routes of Environmental Exposure

The major source of polycarboxylates is household detergents and related cleaning products. For example, in the United States approximately 75% of the household wastewater volume reaches municipal wastewater treatment plants (WWTP), and the remainder is discharged to on-site disposal systems such as septic tanks (Rapaport, 1988). The use of other household cleaning products also leads to discharge of polycarboxylates to wastewater, and, in some cases, to landfills, as a result of wiping surfaces with cloths or paper towels.

Polycarboxylates are consumed in lower market volumes for individual industrial applications than for household applications. However, when they are used in industrial applications, plausible point sources of polycarboxylates include drilling muds for marine applications as well as wastewater from industrial cooling systems.

Mechanisms Affecting Environmental Fate

Table 6.1 provides a summary of ecological fate testing with the above-described 4,500 MW homopolymer and the 70,000 MW copolymer. This testing has indicated that not all polycarboxylates are completely alike in terms of ecological attributes. Although the overall percent removal by wastewater treatment for both polymers will be substantial, the homopolymer and the copolymer differ significantly in terms of fate attributes in wastewater. The following paragraphs briefly describe a comprehensive program to evaluate the ecological fate of polycarboxylates (Rohm and Haas, 1995a,b).

Table 6.1. Summary of polycarboxylate biodegradation and removal in wastewater treatment tests

Test	4,500 MW homopolymer[a]	70,000 MW copolymer[b]
Semi-continuous activated sludge	37.5% removal	94% removal
CO_2 evolution	8.1% CO_2 evolution	20% CO_2 evolution
Batch activated sludge	15.6% CO_2 evolution	17.9% CO_2 evolution
Continuous activated sludge	Typically 50% overall removal with 1 mg/L test concentration but dependent on effluent solids	82% overall removal with 3 mg/L test concentration
Lab scale treatment plant	—	97% overall removal with 5–20 mg/L test concentration
Continuous activated sludge with $FeCl_3$ added	98% removal with 3 mg/L test concentration	—

[a]*Source:* Rohm and Haas, 1995a.
[b]*Source:* ECETOC, 1993.

A practical first step in assessing the ecological fate of polycarboxylates used in laundry detergents is to determine their fate in the wash bath. Removal of the polymers from solution could result if polycarboxylates were adsorbed on fabrics during washing. Solutions containing detergents, polymer, and fabric were agitated at elevated temperatures and aliquots of the solutions removed for polymer analysis. Fabric adsorption of polycarboxylates was not found. Therefore, the key to the ecological fate of polycarboxylates probably lies in removal from waste streams after the use of detergents and other cleaners.

As detergents are primarily discharged to municipal wastewater treatment facilities, it is necessary to focus on removal before and during wastewater treatment. Precipitation of polycarboxylates occurs when carboxyl functionalities in the polymers are neutralized by cationic counterions such as calcium. Once the carboxyl groups have been neutralized, the polymers lose their affinity for water, and precipitation of calcium polycarboxylate takes place. Thus, precipitation can, in some cases, greatly enhance polymer removal from environmental matrices. For example, based on simple laboratory bench tests (adding polymer to tap water and observing precipitate formation) with 30 mg/L 70,000 MW copolymer, calcium polycarboxylate will form in tap water with 200 mg/L hardness as calcium carbonate ($CaCO_3$). However, the precipitation potential is not the same for all polycarboxylates. For 30 mg/L of the 4,500 MW homopolymer, a hardness of 575 mg/L was necessary to cause the calcium polycarboxylate to precipitate. Nevertheless, because wastewater concentrations of polycarboxylate will be much lower than 30 mg/L (see below), there will be sufficient divalent counterions in wastewater to cause at least partial precipitation of the 4,500 MW homopolymer. Figure 6.1 shows the effect of homopolymer concentration on removal by (1) interaction with 200 mg/L hardness tap water alone to determine polymer precipitation with hardness, (2) interaction with 2,500 mg/L autoclaved sludge (suspended in distilled water) from a sewage treatment facility to determine adsorbed polymer, and (3) interaction with a sludge/hard water mixture typical of that found in a treatment plant aeration basin to determine overall removal. Potential biodegradation effects on percent removal were avoided by autoclaving the sludge prior to the experiment.

Precipitation did not fully account for the removal of the homopolymer in the batch removal experiment. The experiment with homopolymer and sewage sludge in deionized water demonstrated that the major portion of polycarboxylate removal results from adsorption onto the sludge. Further, an ecologically important concentration effect on removal was observed. The percent removal increased with decreasing polymer concentration. The homopolymer probably sorbs to the limited number of inorganic sites in sewage sludge, as expected for anionic polycarboxylates. At polymer con-

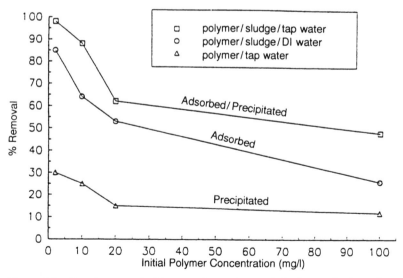

Figure 6.1. Batch removal of homopolymer by precipitation and adsorption.

centrations above approximately 10 mg/L, saturation of available sorption sites significantly diminishes percent removal.

To further investigate the effect of concentration on homopolymer removal, a ^{14}C-labeled 4,500 MW polycarboxylate was used at 0.2 mg/L to 20 mg/L in a batch system containing 2,500 mg/L autoclaved sludge in 175 mg/L hardness water. Figure 6.2 shows the negative correlation of homopolymer removal with concentration. Removal (i.e., precipitation plus adsorption) of 86% was observed for an ecologically relevant initial polymer concentration of 0.2 mg/L.

The batch experiments indicated that at realistic concentrations (typically less than 1 mg/L) the percent removal of the homopolymer will be significantly greater than that predicted with higher polymer concentrations. The high dose batch results were consistent with results from a Semi-Continuous Activated Sludge (SCAS) test, which used 20 mg/L of polymer. The SCAS test demonstrated only 38% removal for the homopolymer compared to 94% removal for the copolymer. The SCAS test incorporates a combination of biodegradation, adsorption, and/or precipitation removal mechanisms. On a mass balance basis, limited adsorption and precipitation of the homopolymer could explain the 38% removal at the high dose of 20 mg/L.

Competing factors that affect polycarboxylate precipitation and adsorption are clearly important in predicting percent removal in wastewater. The calcium stability constant for the homopolymer has been determined to be

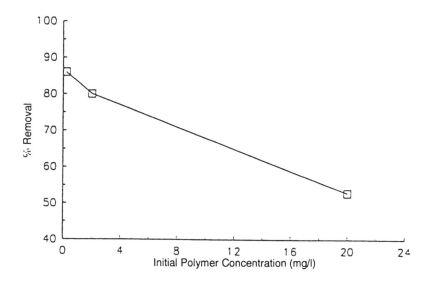

Figure 6.2. Relationship of percent homopolymer removal to concentration.

2.1 by using standard pH titration techniques (Freeman and Bender, 1993). Values of 4.2 and 2.3 were measured for the copolymer and humic acid, respectively. Wastewater components that may compete with polycarboxylates for calcium include residual organic matter and humic substances. These heterogeneous substances can contain relatively high molecular weight polymers that have carboxylic functionalities. Therefore, substances such as humic acid may compete with the homopolymer for calcium.

To further simulate wastewater treatment conditions for polycarboxylates, Continuous Activated Sludge (CAS) tests have been used to evaluate the removal of a ^{14}C-labeled 4,500 MW homopolymer during conventional wastewater treatment. The test system consisted of a mixing chamber in which incoming wastewater and the test substance were mixed before entering the aeration basin. Aliquots of the effluent were fractionated into solid and liquid phases by filtration and assayed to measure homopolymer adsorption. Throughout the entire test, the wastewater and sample feeds were maintained to provide a hydraulic residence time of approximately 6 hr. The test system solids were adjusted to result in a solids retention time of approximately 10 days.

The CAS tests demonstrated an important relationship between wastewater treatment plant conditions and percent removal. Figure 6.3 shows a concentration-dependent relationship between treatment plant effluent

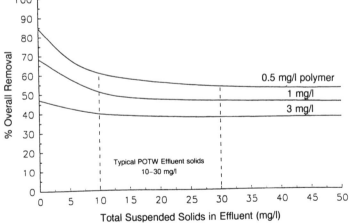

Figure 6.3. Concentration-dependent relationship between treatment plant operation and percent homopolymer removal.

solids and homopolymer removal at several different plant influent concentrations. The CAS tests were run at homopolymer concentrations of 0.5 mg/L to 10 mg/L. The percent removal of homopolymer ranged between 40% to 80%, depending on the level of effluent solids and the polymer concentration. Under typical treatment plant operation and expected influent concentrations of homopolymer (i.e., less than 1 mg/L), 55% homopolymer removal was predicted with the CAS results. Ferric chloride ($FeCl_3$) may be added during primary or tertiary treatment (WPCF, 1990), thereby enhancing the potential for homopolymer removal in some treatment systems. Upon addition of 100 mg/L $FeCl_3$ precipitant to a CAS system to simulate primary or tertiary treatment, virtually complete (i.e., 98%) homopolymer removal was observed. Because homopolymer adsorption to sludge during CAS testing was concentration-dependent (Figure 6.3), the CAS test steady-state adsorption K_d value of 590 for the higher MW fraction can be considered more representative of actual conditions than a previously measured static study K_d value of 2,600 (Hennes, 1991). Although the CAS K_d is lower than the static K_d, the results predict substantial removal of the homopolymer in actual wastewater treatment facilities.

In a study associated with the CAS test, gel permeation chromatography (GPC) was used to evaluate changes in the homopolymer MW distribution (MWD) between the influent and the effluent in the CAS tests (Freeman et al., 1995). Prior to GPC analysis, the CAS effluent samples were clarified by filtration, and a comparison was made of the radioactivity found in the samples after filtration versus that prior to filtration. On average, 33% of the polycarboxylate was removed by filtration. This loss on filtration is indicative of polymer adsorbed on the suspended solids, and further analysis using GPC techniques was used to distinguish the molecular weight dependence of this adsorption. An approximate MWD was calculated for both the filtered CAS effluents and a 4,500 MW homopolymer standard. Figure 6.4 is an overlay of the MWDs that were present in CAS supernatants from experiments with 1, 3, and 10 mg/L homopolymer. The MWDs are almost identical, suggesting that initial polymer concentration does not influence MWD in the effluent. Figure 6.5 shows the MWD of the polycarboxylate in the 1 mg/L CAS test filtered effluent versus the MWD for the dilute polycarboxylate standard, which was similarly fractionated. There was significant loss of the higher molecular weight radiolabeled polycarboxylate due to adsorption on the sewage sludge. These data suggest that the highest molecular weight portions of the polycarboxylate MWD are preferentially adsorbed on the sewage solids.

In addition to removal by adsorption or precipitation, it is important to consider biodegradation of polycarboxylates. Polymers of this type have been shown to have only limited biodegradation, but degradation is an important component of the polymer's overall ecological fate. $^{14}CO_2$ was

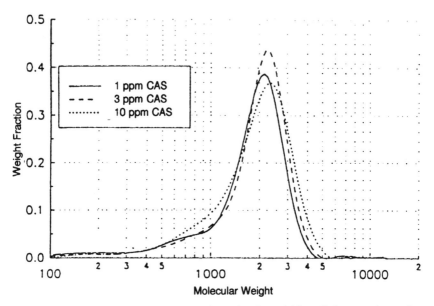

Figure 6.4. Molecular weight distribution of 1, 3, and 10 mg/L homopolymer in CAS effluents.

detected in the air space above the aeration basin in the CAS test containing radiolabeled homopolymer. To verify biodegradation of this polycarboxylate, two separate tests were run. The first was a CO_2 production test, which was coupled with the previously mentioned SCAS test. In this case a microbial inoculum acclimated to the polycarboxylate in the SCAS test was used to rate the extent of the polymer's ultimate biodegradation. This test was run at polycarboxylate concentrations of 10 mg/L and 20 mg/L and demonstrated biodegradation of only 8.1% of theoretical over the course of 28 days. Given the concentration dependence of overall polycarboxylate removal, biodegradation was investigated at 1 mg/L of homopolymer in a $^{14}CO_2$ production test in activated sludge. This test was similar to the previously described CO_2 production test in that an inoculum acclimated in an SCAS test was used to determine the rate and the extent of biodegradation. However, in this case, the level of polycarboxylate (1 mg/L) and the level of biomass (2,500 mg/L) used were typical of those found in a WWTP. As shown in Table 6.1, over the 45-day test period the average total amount of $^{14}CO_2$ evolved was 15.6% of theoretical.

Limited biodegradation of polycarboxylates is typical of many carbon–carbon backbone synthetic polymers. In general, biodegradation of other synthetic polymers may be attributed only to the low molecular weight fractions. Presumably, this is the case for polycarboxylates. Addi-

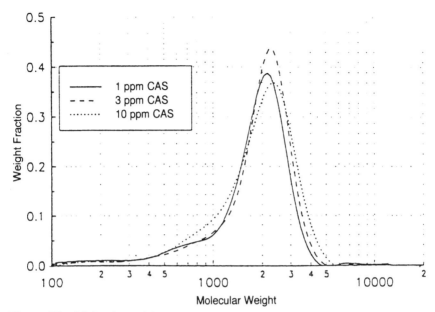

Figure 6.5.   Molecular weight distribution of 1 mg/L homopolymer in a CAS effluent compared to a polycarboxylate standard.

tional research would confirm this with GPC analysis by differentiating between the MWD before biodegradation and the MWD following biodegradation.

Mobility in Sludge and Soil

Mobility studies have demonstrated that once polycarboxylates are adsorbed onto sludge and soil, they are highly unlikely to desorb and leach into aquifers (Chiaudani and Poltronieri, 1990).

Effects on Wastewater Treatment and Metal Remobilization

Supplementary analyses were performed during homopolymer CAS tests (described above) to determine whether the polycarboxylate remobilized metals from the sludge biomass. The samples were analyzed for copper, nickel, cadmium, lead, and zinc during the CAS study. Metal analyses showed that the polycarboxylate had no effect on metal remobilization. Further, addition of the homopolymer at 0.5, 1, 3, and 10 mg/L had no adverse effect on aerobic metabolism within the CAS test system. The total suspended solids and chemical oxygen demand removals of the test and the control units were not significantly different.

Polycarboxylates did not adversely affect the operation of a sewage treatment plant in terms of sludge sedimentation, treatment capacity, and sludge dewatering operations (Chiaudani and Poltronieri, 1990; Opgenorth, 1987, 1992; Schumann, 1991). These results suggest that polycarboxylates do not have adverse effects on the operation of WWTP.

EXPOSURE ASSESSMENT — EXAMPLE

Actual detection and measurement of polycarboxylates in environmental matrices is subject to numerous interferences, principally from the humic acids and related substances found in soils, wastewaters, and surface waters. However, a particularly useful example has used the above ecological fate information to estimate the degree of aquatic exposure to a commercial polycarboxylate found in household laundry detergents (Hamilton et al., 1996). Polycarboxylate-containing detergents are widely distributed to all regions of the United States. Therefore, nationwide polycarboxylate and municipal water consumption can be used to estimate polycarboxylate concentrations that reach WWTP. As noted above, about 26.3 million kg of polycarboxylates was used in detergents and cleaners in the United States in a recent year, primarily as the 4,500 MW homopolymer. The U.S. per capita water use rate is about 400 L/day (U.S. EPA, 1993) and the population approximately 250 million; so, the estimated concentration reaching U.S. sewage treatment plants, based on these data, is 0.7 mg/L. Significant removal of a polycarboxylate will take place before treatment (i.e., by precipitation with calcium in sewer water) and during wastewater treatment (i.e., by biodegradation, precipitation, and adsorption). Further, treated effluent will be diluted upon discharge to natural receiving waters.

Polycarboxylates adsorbed on sewage sludge eventually may be land-filled or used as agricultural fertilizer. In this case, the concentration of polymer in sludge-amended soils can be estimated to be 1.9 mg/kg soil at the time of application (based on an application rate of 2.2 kg dry sludge per square meter of soil per year), with limited biodegradation reducing this concentration over time. The estimated soil concentration level is far below no-effect concentrations in terrestrial test species.

ECOTOXICITY

A substantial amount of ecotoxicity information has been collected with two polycarboxylates, a 4,500 MW homopolymer of acrylic acid and a 70,000 MW copolymer of acrylic acid and maleic acid (Chiaudani and Poltronieri, 1990; Dwyer et al., 1990). The primary research focus has been on the homopolymer and the copolymer because these polycarboxylates are commonly found in widely distributed commercial products. Other polycarboxylates, including those used in industrial water treatment applications, are similar in MW and structure to either the homopolymer or the higher MW copolymer. As indicated below, the ecotoxicity of other polycarboxylates is comparable to that of the homopolymer and the copolymer.

Table 6.2 lists acute and chronic toxicity test results for the homopolymer and the copolymer, indicating that polycarboxylates are probably of "low concern" for acute aquatic and terrestrial toxicity. (A summary of ecotoxicity hazard communication guidelines is presented in Chapter 1.) The 4,500 MW polyacrylic acid homopolymer and the 70,000 MW copolymer have lethal concentrations for 50% of the test population LC_{50}) — or, depending on the test protocol, effect concentrations on 50% of the test population (EC_{50}) — of greater than 100 mg/L of product. Further, the No Observed Effect Concentration (NOEC) of both polycarboxylates is 225 mg/kg or greater of product in a variety of plants and is 1,000 mg/kg or greater of product in earthworms.

Results from chronic toxicity testing with fish consistently suggest that polycarboxylates are of low concern for ecotoxicity in fish, but under some test conditions *Daphnia magna* may be relatively sensitive to chronic exposure. As indicated in Table 6.2, the lowest reported NOEC values for the homopolymer and the copolymer are 12 mg/L and 6.2 mg/L of product, respectively. However, it should be noted that another 21-day chronic toxicity test with the homopolymer has reported a high NOEC of 1,000 mg/L of product.

Both chronic toxicity tests with the homopolymer followed Organisation for Economic Co-operation and Development (OECD) test guidelines. Aquatic toxicity was lower in the test that prevented visible precipitation of polymer to form particulates. Polymer precipitation may have caused adverse external physical effects in *D. magna* such as, but not necessarily limited to, particulate inhibition of respiratory gill function. Toxicity from external physical effects is more plausible than toxicity caused by systemic effects because polycarboxylate uptake by *D. magna* is unlikely. Substances with molecular weights greater than approximately 1,000 have limited permeability potential through biological membranes (Zitko, 1981).

Table 6.2. Ecotoxicity data for household laundry polycarboxylates

Ecotoxity	4,500 MW acrylic acid homopolymer	70,000 MW acrylic acid/maleic acid copolymer
Bacteria		
sludge O_2/glucose consumption (EC_{50})	>100 mg/L	—
Robra O_2 consumption test (EC_{10})	—	>200 mg/L
modified O_2 consumption test (EC_{10})	—	>400 mg/L
Bringmann-Kuehn method (EC_{10})	—	180 mg/L
luminous bacteria method (EC_{20})	—	>200 mg/L
Algae		
Scenedesmus subspicatus (growth inhib., 96 hr EC_{10})	see footnote *d*	>200 mg/L
Selenastrum carpicornutum (growth inhib., 96 hr EC_{50})	—	NOEC = 100,000 mg/L
Hydra		
colony multiplication (EC_{50})	—	136 mg/L
Daphnia magna		
acute (48 hr, immobilization, EC_{50})	>1,000 mg/L	>200 mg/L
chronic reproductive (21 day, NOEC)	12 mg/L; 1,000 mg/L	6.2 mg/L
Fish		
Acute:		
rainbow trout (96 hr, LC_{50})	700 mg/L	—
bluegill sunfish (96 hr, LC_{50})	>1,000 mg/L	—

zebra fish (96 hr, LC_{50})	200 mg/L	—
golden orfe (96 hr, LC_{50})	—	>200 mg/L
Chronic:		
Hydra littoralis (14 day EC_{10} and EC_{50})	—	40 mg/L EC_{10} and 136 mg/L EC_{50}
larval test (zebra fish, 6 weeks, NOEC)	—	40 mg/L
sublethal test (zebra fish, 14 day, NOEC)	—	40 mg/L
early life stage (zebra fish, 28 day, NOEC)	450 mg/L	—
early life stage (fathead minnow, 21 day, NOEC)	124 mg/L	—
Plant		
corn, wheat, soybean (growth inhib., NOEC)	225 mg/kg soil	—
oat (growth inhib., NOEC)	—	400 mg/kg soil
Earthworm		
acute (96 hr, LC_{50})	>1,000 mg/kg soil	>1,600 mg/kg soil

[a] All results reported here according to mg/L or mg/kg of *product*. A typical polycarboxylate product consists of approximately 45% by weight polymer active (includes 2% reaction by-products) and 55% water.

[b] *Sources:* Rohm and Haas, 1995a; Hennes, 1991.

[c] *Source:* ECETOC, 1993.

[d] *Scenedesmus suspicatus* 96 hr, EC_{10} = 180 mg/L (ECETOC, 1993), but water hardness level was not reported. See text for more details on algae toxicity.

Based on a 96-hr EC_{10} for growth inhibition of 180 mg/L of product (see Table 6.2), algae (*Scenedesmus subspicatus*) do not appear to be markedly sensitive to the homopolymer. However, the reference does not cite the test conditions, and work on related polycarboxylates demonstrates that factors such as the water hardness greatly impact the results. Hardness levels of 15 mg/L to 20 mg/L as $CaCO_3$ are typically used in the growth medium in standard algal toxicity tests. Higher hardness levels can mitigate the toxicity of polycarboxylates to algae. For example, the 96-hr EC_{50} in green algae with an acrylic acid homopolymer has been shown to change from an EC_{50} of 37 mg/L (as polymer) at 15 mg/L hardness to an EC_{50} of 780 mg/L (as polymer) at 158 mg/L hardness (Nabholz and Zeeman, 1994). The higher hardness is typical of the water hardness in the United States (e.g., 120 mg/L). The apparent toxicity at 15 mg/L hardness may be associated with excess sequestration of essential cationic nutrients by the polycarboxylates. The presence of additional calcium may have moderated the sequestration of nutrient elements in the growth medium. Overall, algae do not appear to be sensitive to polycarboxylates under realistic hardness conditions.

Polycarboxylates have the potential to have important physical-chemical interactions with test species, particularly at high doses during laboratory testing. In a series of recent studies, an interesting relationship between polycarboxylate dose and developmental effects with sea urchins was observed. A 2,100 MW acrylic acid homopolymer was tested for effects in vitro on sea urchin sperm respiration and embryo development (Cekolin et al., 1993) and for effects in vitro on sea urchin fertilization (Cekolin, 1991). No polymer-related effects on fertilization and sperm respiration were observed with up to 1,000 mg/L polymer when applied to sea urchin sperm, but the polymer reduced embryo respiration at 1,000 mg/L. Exposure to the polymer produced embryos with abnormal shell spicules at 100 mg/L, and when the polymer was applied specifically during the blastula stage of embryonic development, shell spicule development was inhibited at 1,000 mg/L. This is not surprising because polycarboxylates are designed to inhibit mineralization (i.e., precipitation) of calcium salts. The sea urchin rapidly develops from a zygote to a shelled free-swimming organism (pluteus) with a calcium-containing shell within about 40 hours (Cekolin, 1991). Therefore, the effects observed in vitro were consistent with exposure to very high (i.e., 100 mg/L and above) concentrations of polycarboxylates.

Other polycarboxylates show a comparable ecotoxicity profile to that of the 4,500 MW homopolymer and the 70,000 MW copolymer. As shown in Table 6.3, polycarboxylate copolymers and terpolymers based on acrylic acid, methacrylic acid, sulfonic acids, and/or acrylamides are of low concern for aquatic toxicity. Normal levels of water hardness (i.e., 100 mg/L hardness or greater) significantly mitigated toxicity of the acrylic acid–methacrylic acid copolymer in algae.

Table 6.3. Aquatic toxicity of other polycarboxylates

Ecotoxicity	3,500 MW acrylic acid–methacrylic acid[a,b] (mg/L)	4,500 MW acrylic acid–AMPS[b,c] (mg/L)
Algae (growth inhib., 96 hr EC$_{50}$)		
▪ nonmitigated test medium (15 mg/L hardness)	6.3	58
▪ calcium added (1:1 calcium–polymer ratio)	58	—
▪ test medium hardness 50 mg/L	35	—
▪ test medium hardness 100	210	—
▪ test medium hardness 150	990	—
Daphnia magna		
▪ acute (48 hr, immobilization, EC$_{50}$)	>1,000	>1,000
Fish		
▪ rainbow trout, (96 hr, LC$_{50}$)	>1,000	>1,000
▪ bluegill sunfish, (96 hr, LC$_{50}$)	>1,000	>1,000

[a]All results reported here according to mg/L or mg/kg of *product*. A typical polycarboxylate product consists of 45% polymer (includes 2% reaction by-products) and 55% water.

[b]*Sources*: Rohm and Haas, 1995b; Hennes, 1991.

[c]AMPS = acrylamidomethylpropane sulfonic acid.

Conclusions

Polycarboxylates are of low overall concern for ecotoxicity and appear to be used at ecologically responsible levels. Although two commercially important polycarboxylates have chronic NOEC values that suggest moderate concern for chronic toxicity in *D. magna*, actual polycarboxylate concentrations in surface waters will be substantially below no-effect concentrations. Further, the ecotoxicity profile of polycarboxylates suggests low concern for adverse effects in earthworms and plants. Water hardness can significantly mitigate the toxicity of polycarboxylates in algae. Ecological fate studies have shown that it is critical to test at or near ecologically realistic exposure levels, and that polycarboxylates in wastewater will be significantly removed before and during wastewater treatment by a combination of biodegradation, adsorption, and precipitation.

Part II
Polyacrylate Superabsorbents

■ SERGIO S. CUTIÉ AND STANLEY J. GONSIOR
  The Dow Chemical Company

INTRODUCTION

Polyacrylate superabsorbents (PAS) are acrylic polymers that can absorb and retain many times their weight of water or aqueous solutions. Their unique water absorption properties make them particularly suitable for use in disposable products such as diapers, adult incontinence products, and female hygiene pads. For example, the fluid absorption standard for commercial PAS is between 30 g and 40 g of urine absorbed per g of dry PAS. PAS are prepared by polymerizing acrylic acid, sodium acrylate, and a multifunctional cross-linking agent (0.001–1% by weight) in water to form a three-dimensional network with infinite molecular weight. Because of this cross-linking, PAS are virtually insoluble in water apart from a small amount of non-cross-linked polymer (<10%) and traces of residual monomer. Commercial PAS is a granular solid, but when used to absorb fluid in an absorbent product, the PAS matrix will swell to form a gel.

Following their use, most products containing PAS enter the municipal solid waste (MSW) stream. Currently, the disposal of MSW in the United States involves landfilling (62%), incineration (16%), recycling (18%), and composting (4%) (U.S. EPA, 1994). In addition, a small portion of PAS may enter sewage effluents following inadvertent flushing of hygiene products into the sewer system. Therefore, the ecological fate of PAS in landfills, composting, and wastewater treatment plants has been investigated. In addition, the fate of PAS, including water-soluble fractions of PAS, has been evaluated in soil column studies to complement the studies involving landfills.

ECOLOGICAL FATE

Analytical Methods

Monitoring the fate of PAS in environmental studies presents a challenge, given the required sensitivity and specificity for measuring concentrations of PAS in complex environmental matrices such as soil. The insoluble, cross-linked polymer network of PAS makes routine chromatographic procedures impossible. To characterize PAS with hydrolyzable ester cross-links, it is possible first to hydrolyze the cross-link network under heat and high pH, and then to use size exclusion chromatography to determine molecular weight distributions of PAS samples (Cutié and Martin, 1995). However, not all commercial PAS have ester cross-links.

Several analytical methods have been developed to quantify PAS in a sample by measuring the concentration of sodium present. A sodium ion-selective electrode was successfully used to measure the concentration of sodium polyacrylates and their percent neutralization in certain matrices (Cutié et al., 1992). A procedure was developed to quantify PAS dust in air by measuring sodium, using either atomic absorption or atomic emission spectroscopy (Forshey et al., 1994). Procedures that rely on the measurement of sodium require close attention to sample collection, preparation, analysis, and data interpretation because of the potential for contamination by background sodium. Recently, a very sensitive and specific procedure to determine PAS dust in air was developed, which utilizes the inherent ion exchange properties of PAS and the sensitivity of neutron activation analysis (Rigot et al., 1996). This procedure exchanges the sodium present in the PAS for europium, thereby eliminating interferences from sodium and similar background ions.

PAS contain a small fraction of non-cross-linked, polydisperse polymer with molecular weights ranging from $> 1,000$ to $< 1,000,000$. This fraction, containing primarily polyacrylates, is soluble in water and thus potentially more susceptible to environmental processes such as movement in soil and biodegradation than the cross-linked polymer fraction. Several analytical methods have been developed to measure the concentration of polyacrylates in environmental samples. Pyrolysis-gas chromatography coupled with detection by either mass spectrometry, flame ionization detection, or infrared spectroscopy has been successful in certain applications (Buzanowski et al., 1994). However, pyrolysis of aqueous extracts from such matrices as soil produces interferences that make quantitation of the polyacrylate difficult, particularly at ecologically relevant (i.e., trace) concentrations. High performance gel permeation chromatography with refractive index detection was developed to quantify polyacrylates in the eluents from soil column studies using sandy soil (Cutié et al., 1990; Sack et al., 1996). However, with soils having a high organic matter content, co-elution of interferences reduces the analytical sensitivity of the method.

When possible, ecological fate studies for PAS have been conducted by using radiolabeled PAS, typically incorporating ^{14}C at the 2- and 3-carbon positions on the polymer backbone (Martin et al., 1987a, b, 1990; Pohland et al., 1993; Rittman et al., 1992a, b; Stegmann et al., 1993). The use of a radiotracer, coupled with sensitive radio-analytical techniques, has facilitated the determination of the fate of PAS in complex matrices ranging from landfills to soil columns. Detection of $^{14}CO_2$ or $^{14}CH_4$ in gaseous emissions in several studies demonstrated some mineralization of the polymer backbone. However, complete characterization of ^{14}C-material in leachates and effluents from these studies (e.g., molecular weight) was not performed.

Landfills

Landfills are the predominant route of disposal for MSW in the United States. Long-term studies have elucidated the fate of PAS in simulated landfills. ^{14}C-PAS was incorporated into diapers that were added at realistic levels (approximately 2% dry weight) to shredded MSW (Pohland et al., 1993; Stegmann et al., 1993). The MSW was then added to gastight lysimeters (120–2,100-liter capacity) configured to allow leachate recirculation. Landfill stabilization typically occurs over decades and involves progression through several phases including initial adjustment, transition, acid formation, methane formation, and final maturation (Pohland et al., 1993). In the simulated landfill studies, temperature control and leachate recycle were used to accelerate the stabilization process. In a 4-year study, less than 3% of the total radioactivity from ^{14}C-PAS was found in effluent gas (as $^{14}CO_2$ or $^{14}CH_4$) and leachate from the simulated landfill. In a 1.5-year study, radioactivity measured in the effluent gas was less than 1%, and 2% to 4% was found in the leachate. Most of the ^{14}C-PAS (>90%), remained in the diapers and nearby surrounding waste. Compared to results from control lysimeters (i.e., no PAS added), no adverse effects on landfill processes (measured by gas generation) or leachate quality (measured by chemical oxygen demand, biochemical oxygen demand, volatile acids, and pH) were observed in the 1.5-year and 4-year studies.

Results from the landfill studies support biodegradation of the small portion of non-cross-linked, lower molecular weight polymer present in the PAS. Aerobic biodegradation can be operative in the initial adjustment phase of a landfill stabilization process. As the stabilization process proceeds, anaerobic biodegradation processes typically predominate. In the 1.5-year study, ^{14}C-material in the leachate was determined to be highly biodegradable aerobically. In addition, some anaerobic biodegradation of an isolated fraction (MW 16,700) of the non-cross-linked portion of PAS was demonstrated elsewhere (Rittmann et al., 1992a). These results are consistent with the observed biodegradability of similar ^{14}C-polycarboxylates with molecular weights ranging from 1,000 to 10,000 (Chiaudani and Poltronieri, 1990).

Composting

The fate of PAS in compost has been determined under aerobic composting conditions (Stegmann et al., 1993). ^{14}C-PAS was incorporated into diapers added to MSW (approximately 2% dry weight) and composted in a gastight lysimeter (120-liter volume) for 510 days. ^{14}CO$_2$ in the effluent gas (6% of the total activity added) indicated that some mineralization of the ^{14}C-PAS occurred. As noted previously, the lower molecular weight polymers present in the PAS were most likely biodegraded. Little radioactivity was found in the leachate (0.6%), indicating minimal leaching of ^{14}C-material out of the compost. As found in the landfill studies, diapers containing PAS had no adverse effect on the composting process in terms of chemistry of the leachate and gas production.

Soil Columns

To complement research on the fate of PAS in landfills, the mobility of PAS in soil was addressed in several soil column studies (Martin et al., 1987a, 1990b; Rittmann et al., 1992b; Sack et al., 1996). Most studies used a sandy soil and artificially high loadings of polymer to simulate a "worst-case" condition, that is, one having the greatest potential for polymer migration. The low organic test soil (Borden sand, 96% sand, 2% silt, 2% clay, <0.5% organic matter) was taken from an experimental field site in Canada. The initial form of the PAS added to the soil columns varied, including granular polymer, polymer incorporated into cellulose pads and filtered aqueous polymer suspensions, as well as aqueous suspensions of the soluble fraction of PAS fractionated on the basis of molecular weight. In addition, 4,500 MW sodium polyacrylate was included to represent the small portion of PAS expected to be most susceptible to movement in soil. This polyacrylate was an acrylic acid homopolymer (i.e., polycarboxylate). Various solutions, ranging from distilled water to synthetic groundwaters or landfill leachates, were passed through the columns to elute the added polymers. A summary of results from the column studies is shown in Table 6.4.

Table 6.4. Mobility of SAP and its water-soluble fractions in sand columns

Polymer form[a]	Aqueous medium[b]	Column length (cm)	Pore volumes passed	% Polymer eluted from column	Reference
^{14}C-PAS in cellulose pad	Distilled water	35	>9	4.6	Martin et al., 1987a
"	Sterilized leachate	35	>9	5.6	Martin et al., 1987a
PAS	Synthetic groundwater	30	30	<1	Sack et al., 1996
Fractional MW 16,700 from ^{14}C-PAS	Synthetic groundwater	3	5.8	1.2	Rittmann et al., 1992b
Polyacrylate MW = 4,500	Synthetic groundwater	30	30	<8	Sack et al., 1996
^{14}C-Polyacrylate MW = 4,500	Synthetic groundwater	10	107	9.5	Chiaudani and Poltronieri, 1990
Filtered solution from ^{14}C-PAS[c]	Distilled water	20	2.9	41.6	Martin et al., 1987a
"	Distilled water	20	47.3	57.7	Martin et al., 1987a
"	Distilled water	20	97.5	52.2	Martin et al., 1987a
"	NaCl solution	20	43.7	45.5	Martin et al., 1987a
"	Sterilized leachate	20	46.4	32.0	Martin et al., 1987a

[a]Form of polymer added to sand column.

[b]Type of aqueous medium passed through column.

[c]Aqueous suspension of ^{14}C-PAS was filtered, presumably removing insoluble, cross-linked polymer. ^{14}C-material in filtrate was not characterized.

Introduction of ^{14}C-PAS to a sand column as part of a cellulose pad resulted in only about 5% of the radioactivity migrating out of the column (Martin et al., 1987a). In a similar study, less than 1% of the PAS eluted from the column as polyacrylate, as determined by size exclusion chromatography with refractive index detection (Sack et al., 1996). In both cases, most of the PAS introduced into the columns was in the form of immobile, cross-linked polymer. Other column studies focused on the small portion of PAS that is water-soluble. A fraction of non-cross-linked polymer with MW 16,700, isolated from ^{14}C-PAS by fraction precipitation, was added to a sand column (Rittmann et al., 1992b). In this case, only 1% of the radioactivity eluted from the column, while 98% remained in the first quarter of the column. For the low molecular weight (4,500 MW) sodium polyacrylate, less than 10% of the added polymer eluted from sand columns in separate studies (Chiaudani and Poltronieri, 1990; Sack et al., 1996). Introduction of a filtrate of an aqueous suspension of ^{14}C-PAS to sand columns resulted in 32% to 58% elution of ^{14}C-material from the columns after passage of 3 to 50 pore volumes of aqueous solution (Martin et al., 1987a). Movement of the remainder of the radioactivity was highly retarded, and most remained near the column inlet. Characterization of the ^{14}C-material present in the filtrate and the column eluent was not performed. However, these results suggest that most of the cross-linked polymer in the aqueous suspension of the ^{14}C-PAS was removed during prior filtration, leaving primarily water-soluble components to be evaluated in the study.

To expand upon the soil column studies and provide a three-dimensional simulation of an aquifer environment, ^{14}C-PAS was introduced as part of a cellulose pad into an aquifer simulation tank containing sand (Martin et al., 1987b). Following the passage of six pore volumes of groundwater, 1.9% to 3.5% of the radioactivity was discharged from the tank. These results, together with the results from the various column studies, suggest that only a small portion of the 60% neutralized water-soluble fraction of PAS, primarily low molecular weight polymer, was susceptible to movement through sand. These results are consistent with the behavior of nonneutralized polycarboxylates, where mobility in soil decreases with increasing molecular weight (Chiaudani and Poltronieri, 1990).

Wastewater Treatment Plants

The fate of PAS was assessed in a pilot-scale aerobic wastewater treatment plant containing primary settling, aeration, and clarifying chambers (Martin et al., 1990). A control plant was fed predominantly domestic wastewater amended with cellulose fiber products at a maximum anticipated usage rate. A test plant received the same influent, except that PAS was added to the fiber product. After 32 weeks of operation, ^{14}C-PAS was added to the influent of the test plant. Most of the polymer (>97%) was found to be associated with solids deposited in the primary settling and aeration chambers with minimal releases in the effluent. Operation of the test plant was essentially the same as that of the control plant, indicating no adverse effect of the PAS on the treatment plant's various treatment processes.

CONCLUSIONS

Available ecological fate studies represent a comprehensive effort to define the fate of PAS in consumer products in various routes of disposal, including landfills, composting operations, and wastewater treatment plants. PAS had no adverse effects on the biological processes occurring in pilot- and large-scale simulations of those disposal systems. In addition, little of the PAS and its degradation products was found in gas emissions, leachate, or effluent.

Although ecotoxicity information is not yet available in the published literature on PAS alone, information on a water-soluble 4,500 MW polycarboxylate (see Part I of this chapter) that was used in PAS soil column studies suggests very low concern for ecological effects. This low molecular weight polycarboxylate was used in the soil column studies to represent the small portion of the water-soluble fraction of PAS. Further, the soil column studies simulated worst-case conditions for potential polymer migration by using high PAS loadings in sandy soil. The cross-linked portion of PAS was virtually immobile in sand, whereas the small water-soluble fraction had limited susceptibility to movement. Therefore, under realistic conditions in landfills, composts, and soils, the mobility of PAS will be negligible.

REFERENCES (PARTS I & II)

Buzanowski, W. C., S. S. Cutié, R. Howell, R. Papenfuss, and C. G. Smith. 1994. Determination of sodium polyacrylate by pyrolysis-gas chromatography. *Journal of Chromatography* A677:355–64.
Cekolin, C. S. 1991. Toxicity studies of antiscalant agents using *Arbacia punctulata* gametes and embryos as test organisms. M.S. thesis. University of South Alabama, Department of Biological Sciences, Mobile, AL.

Cekolin, C. S., J. E. Donachy, and C. S. Sikes. 1993. Toxicity studies of antiscalant agents using *Arbacia punctulata* gametes and embryos as test organisms. *Bulletin of Environment Contamination and Toxicology* 50: 108–15.

Chiaudani, G. and P. Poltronieri. 1990. Study on the environmental compatibility of polycarboxylates used in detergent formulations. *Ingegneria Ambientale* 11:1–43.

Cutié, S. S. and S. J. Martin. 1995. Size-exclusion chromatography of cross-linked superabsorbent polymers. *Journal of Applied Polymer Science* 55:605–9.

Cutié, S. S., W. C. Buzanowski, and J. A. Berdasco. 1990. Fate of superabsorbents in the environment, analytical techniques. *Journal of Chromatography* 513:93–105.

Cutié, S. S., R. M. Van Effen, D. L. Rick, and B. J. Duchane. 1992. Sodium ion-selective electrode to determine superabsorbent polymers and to determine their degree of neutralization. *Analytica Chimica Acta* 260:13–17.

Dwyer, M., S. Yeoman, J. N. Jester, and R. Perry. 1990. A review of proposed non-phosphate detergent builders, utilization, and environmental assessment. *Environmental Technology* 11:263–94.

ECETOC. 1993. *Joint Assessment of Commodity Chemicals: Polycarboxylate Polymers*, Report No. 23. Brussels, Belgium: European Chemical Industry Ecology and Toxicology Centre.

Forshey, P. A., T. S. Turan, J. S. Lemmo, S. S. Cutié, and D. L. Pytynia. 1994. Analysis of sodium polyacrylate absorbent dust using ultra-trace sodium analysis — a seven company collaborative study. *Analytica Chimica Acta* 298:351–61.

Freeman, M. B. and T. M. Bender. 1993. An environmental fate and safety assessment for a low molecular weight polyacrylate detergent additive. *Environmental Technology* 14:101–12.

Freeman, M. B., T. M. Bender, and J. D. Hamilton. 1995. The Impact of a Polymer's Molecular Weight Distribution on Risk Assessment, paper presented at the 15th Annual Meeting of the Society of Environmental Toxicology and Chemistry, Denver, CO.

Hamilton, J. D., M. B. Freeman, and K. H. Reinert. 1996. Aquatic risk assessment of a polycarboxylate polymer used in laundry detergents. *Journal of Toxicology and Environmental Health* (accepted for publication).

Hennes, E. C. 1991. Fate and Effects of Polycarboxylates in the Environment. Technical Report. Strombeek-Bever, Belgium: Procter & Gamble Company.

Martin J. E., K. W. F. Howard, and L. W. King. 1987a. Environmental behavior of ^{14}C-tagged polyacrylate polymer: Column studies of flow and retardation in sand. *Nuclear and Chemical Waste Management* 7: 265–71.

Martin, J. E., K. W. F. Howard, and L. W. King. 1987b. Environmental behavior of ^{14}C-tagged polyacrylate polymer: Flow tank studies of retention in sand. *Nuclear and Chemical Waste Management* 7:273–80.

Martin, J. E., T. Stevens, G. E. Bellen, L. W. King, and J. M. Hylko. 1990. Carbon-14 tracer study of polyacrylate polymer in a wastewater plant. *Applied Radiation Isotope* 41:1165–72.

Nabholz, J. V. and M. G. Zeeman. 1994. The Environmental Concerns of Polymers from the Perspective of OPPT/USEPA, paper presented at the 15th Annual Meeting of the Society of Environmental Toxicology and Chemistry, Denver, CO.

Opgenorth, H. J. 1987. Environmental tolerance of polycarboxylates. *Tenside Surfactants Detergents*, 24:366–69.

Opgenorth, H. J. 1992. Polymeric materials polycarboxylates, in *The Handbook of Environmental Chemistry,* Vol. 3, Part F, ed. O. Hutzinger. New York: Springer-Verlag.

Pohland, F. G., W. H. Cross, and L. W. King. 1993. Codisposal of disposable diapers with shredded municipal refuse in simulated landfills. *Water Science Technology* 27(2):209–23.

Rapaport, R. A. 1988. Prediction of consumer product chemical concentrations as a function of publically owned treatment works treatment type and riverine dilution. *Environmental Toxicology and Chemistry* 7:108–15.

Rigot, W. L., H. W. Emmel, and S. S. Cutié. 1996. Submitted for publication in *Analytica Chimica Acta.*

Rittmann, B. E., B. Henry, J. E. Odencrantz, and J. A. Sutfin. 1992a. Biological fate of a polydisperse acrylate polymer in anaerobic sand-medium transport. *Biodegradation* 2:171–79.

Rittmann, B. E., J. A. Sutfin, and B. Henry. 1992b. Biodegradation and sorption properties of polydisperse acrylate polymers. *Biodegradation* 2:181–91.

Rohm and Haas Company. 1995a. *Formulation Chemicals Research: An Environmental Safety Assessment of Acusol® 445N.* Technical Report. Spring House, PA: Rohm and Haas Company.

Rohm and Haas Company. 1995b. *Formulation Chemicals Research: Summary of Environmental Toxicity Tests of Polycarboxylates.* Technical Summary. Spring House, PA: Rohm and Haas Company.

Sack, T. M., J. Wilner, S. S. Cutié, and A. R. Blanchette. 1996. Submitted for publication in *Environmental Science and Technology.*

Schumann, H. 1991. Elimination properties of polyelectrolytes in biological waste water purification processes. *Tenside Surfactant Detergents* 28:452–59.

SRI. 1992. *Water Soluble Polymers.* Specialty Chemicals Report, San Francisco, CA: Stanford Research Institute.

Stegmann, R., S. Lotter, L. King, and W. D. Hopping. 1993. Fate of an absorbent gelling material for hygiene paper products in landfill and composting. *Waste Management Research* 11:155–70.

U.S. EPA. 1993. Standards for the Use or Disposal of Sewage Sludge. U.S. Government Code of Federal Regulations. 40 CFR Part 257, pp. 9248–415. Washington, DC: United States Environmental Protection Agency.

U.S. EPA. 1994. *Characterization of Municipal Solid Waste in the United States.* 1994 Update, EPA 530-S-94-042. Washington, DC: United States Environmental Protection Agency.

WPCF. 1990. *Operation of Municipal Waste Water Treatment Plants.* Alexandria, VA: Water Pollution Control Federation.

Zitko, V. 1981. Uptake and excretion of chemicals by aquatic fauna, in *Ecotoxicology and the Aquatic Environment*, ed. P. Stokes. New York: Pergamon Press.

Chapter 7

Water Treatment Polymers

■ LARRY A. LYONS AND STEPHEN R. VASCONCELLOS
BetzDearborn, Inc.

INTRODUCTION

The use of synthetic polymers has contributed to significant advancements in the technology of wastewater and supply water treatment. Water treatment polymers are used in a wide range of industries, including refineries (e.g., oil, chemical), pulp and paper mills, and all types of manufacturing facilities (e.g., steel, automotive, food processing, textile), as well as in municipal wastewater treatment plants. Water treatment processes convert liquid wastes into acceptable final effluents, and water treatment polymers allow the effective disposal of solids and other contaminants that are received by or generated from treatment processes.

Treatment is often required for insoluble, suspended, and dissolved contaminants. Wastewater can include a wide range of organic and inorganic compounds, nutrients (i.e., nitrogen and phosphorus), and metals. Water-soluble contaminants are converted chemically or undergo biological oxidation during treatment to form insoluble solids. Also many insoluble materials are generated from manufacturing and treatment processes (e.g., oils, organics, greases, paints, and inorganic particulates). Water for industrial or domestic use can be treated with polymers to reduce color caused by organics such as lignins and humic acids, and to remove other suspended and dissolved contaminants.

Water treatment polymers are high molecular weight structures that react with colloidal and suspended substances to create removable solids, often referred to as floc, which can be effectively separated from the water being treated. Table 7.1 presents typical polymer applications for water

Table 7.1. Types of water treatment polymer applications

Clarification of influent water: Removing suspended solids and colloidal organics that cause color from river or lake water. This application is necessary for many industrial processes.

Primary and secondary clarification: Removal of suspended solids, oils, and organic contaminants for clarification of wastewater.

Flotation: Coagulation for flotation or separation of oils from water.

Thickening: Flocculation of suspended solids to concentrate solids, often for sludge waste-handling operations.

Dewatering: Flocculation to remove water from sludge slurry to further concentrate for solid waste disposal. Solid contents of 1–4% often can be concentrated to 10–40% per unit volume.

Tertiary: Polishing applications such as metals removal.

---

treatment. A primary benefit of polymer applications is efficient performance under high hydraulic volume and throughput conditions. For instance, the throughput of industrial wastewater can easily exceed 2 million gallons per day; and a municipal treatment plant that serves a population of 100,000 is often designed to process wastewater streams with a throughput greater than 50 million gallons per day. Polymers also offer other advantages, particularly when compared to inorganic coagulant aids (i.e., alum or ferric chloride), such as minimizing or reducing sludge volumes, being noncorrosive to treatment systems, minimizing chemically bound water, providing no additional contribution to total dissolved solids concentrations, requiring lower dosages, and usually not requiring additional treatment (e.g., pH adjustments).

ENVIRONMENTAL OVERVIEW

Application of water treatment polymers provides important environmental pollution control and water reuse benefits by clarification and removal of solids, reduction or elimination of many kinds of contaminants, and consolidation of sludge for solid waste disposal. However, an understanding of (1) polymer partitioning characteristics between liquid (i.e., soluble) and bound (i.e., solid) phases and (2) ecological effects expressed by liquid and solid phases is necessary for an assessment of ecological risks associated with water treatment polymers. The bioavailability and the toxicity of water treatment polymers to aquatic organisms are highly dependent on the degree of partitioning between liquid and solid phases. Ecological effects can be significantly mitigated when water treatment polymers are bound to organics. Further, of particular ecological relevance are the polymer application procedures used in wastewater treatment systems. When followed properly, these procedures can optimize the adsorption of polymers and minimize the amount of residual (i.e., free) polymer in the liquid phase. Selecting proper polymers, optimally administering polymers to the treatment system, and adjusting polymer feed for wastewater variability can effectively control the aquatic toxicity and the fate of water treatment polymers.

POLYMER CLASSIFICATION

General Aspects

Water treatment polymers are predominantly water-soluble, and are classified by their ionic charge potential as cationic, anionic, nonionic, or amphoteric. Water solubility and charge are imparted by incorporating ionizable sites along the polymer backbones. Those polymers with a positive charge are cationic, and those with a negative charge are anionic. Water-soluble polymers with no ionizable groups are considered to be nonionic. A fourth minor class includes amphoteric polymers, which are zwitterionic polymers having both cationic and anionic functional groups. The nature of the charge for the amphoterics is dependent upon the pH of the wastewater.

Water treatment polymers are further classified into two main groups: coagulants and flocculants. Both coagulants and flocculants react with colloidal material by neutralizing the charge and/or chemically bridging individual particles to form a visible insoluble floc (Figures 7.1 and 7.2). Polymers that function primarily by charge neutralization are considered to be coagulants, whereas polymers that remove particulates by both charge neutralization and bridging are flocculants. These two groups of polymers can be further distinguished by the degree of charge and their molecular weight, with flocculants being much higher in molecular weight and having a broader range of ionicity than coagulants. Molecular structures of some water treatment polymers can be found in Chapter 1. Table 7.2 presents a general classification of polymeric coagulants and flocculants. The ionicity and the charge density of water treatment polymers both refer to the

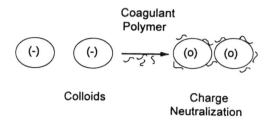

Figure 7.1. Coagulation mechanism.

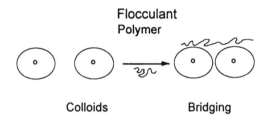

Figure 7.2. Flocculation mechanism.

fraction of ionizable groups along the polymer backbone. Ionicity is given as mole percent charge. Charge density relates ionicity to millequivalents per gram (meq/g).

Polymeric Coagulants

Polymeric coagulants are cationic polyelectrolytes, which can be either polyamines (R2NH) or quaternary amines ($R4N^+$). As polyelectrolytes in solution they can exhibit strong cationic ionization. In water, polyamines react as follows:

$$R2NH + H_2O \rightarrow R2NH{-}H^+ + OH^-$$

Because this reaction is pH-dependent, nonquaternized polyamines (secondary and tertiary amines) will show a decrease in cationicity with increasing pH. Cationic charge densities of polyamines, measured by colloid titration, ranges from 4 to 10 meq/g, at acidic (pH < 4) conditions. In contrast, quaternary polymers are only slightly affected by pH, remaining positively charged over a broad pH range. The quaternary amine functionality does not include hydrogen ions:

$$R4N^+ + H_2O \rightleftarrows R4N^+ + H_2O$$

Table 7.2. Classification of water treatment polymers

I. Coagulants
   A. Quaternized cationic monomers–homopolymers and copolymer with acrylamide
      1. Epichlorohydrin/dimethylamine (EPI/DMA)
      2. Diallydimethylamine (DADMAC)
   B Nonquaternized low molecular weight cationic polymers
      1. Polyamines
      2. Polyethyleneimines/polyvinylamines
   C. Cyclic cationic polymers
      1. Melamine-formaldehydes (MF)
      2. Pyridine derivatives
      3. Other

II. Flocculants
   A. Cationic polymers
      1. Cationic monomers with acrylamide (AM)
         a. Esters
         b. Amides
      2. Nonquaternized high molecular weight cationic polymers
         a. Mannich polyacrylamides
   B. Anionic polymers
      1. Anionic monomers with AM
         a. Acrylic acids (e.g., acrylic acid, abbrev. AA)
         b. Mineral acids (e.g., acrylamidomethylpropane sulfonic acid, abbrev. AMPS)
   C. Nonionic polymers
      1. Polyacrylamides (PA)
      2. Polyethylene glycols
   D. Amphoterics

The cationic charge density for quaternized polymers ranges from 5 meq/g to 10 meq/g. Both polyamines and quaternary polymers typically have molecular weights of less than 500,000.

Polymeric Flocculants

Flocculant polyelectrolytes have much higher molecular weights than primary coagulants and a broader range of charge, and can be anionic, nonionic, or cationic. Typical flocculants are polyacrylamide-based copolymers. These copolymers contain cationic or anionic functional monomers to vary the degree of ionicity. Molecular weights of these copolymers can range from 1 million to greater than 50 million. Flocculants function by both charge neutralization and bridging; in bridging, long bridges between small flocs are formed, enhancing particle growth. Charge densities will range from <1 meq/g to 5 meq/g.

Anionic polymers incorporate carboxyl groups (—COOH) in their structure. They ionize as follows:

$$R{\rm —COOH} \rightleftharpoons R{\rm —COO}^- + H^+$$

The hydrogen ion forces the reaction to the left, so anionics become nonionic at low pH. Anionic flocculants are used as flocculant aids in influent clarification processes to promote settling, and in clarification and dewatering of inorganics sludges. They can also be used with alum for sludge dewatering in paper production effluent treatment processes.

Nonionic water treatment polymers specifically contain acrylamide as the monomeric building block. These polymers can function as effective flocculants when polar sites on the particulate molecule dominate the destabilization mechanism. Nonionic polyacrylamides can be effective in attracting and holding colloidal particles at the polar sites.

Cationic polyacrylamides are composed of acrylamide and cationic monomers such as acryloxloxyethyl trimethyl ammonium chloride (AETAC), methacryloyloxy ethyltrimethyl ammonium chloride (METAC), acrylomidopropyl trimethyl ammonium chloride (APTAC), methacrylomidopropyl trimethyl ammonium chloride (MAPTAC), and/or diallyldimethyl ammonium chloride (DADMAC). These flocculants are typically effective over a pH range of 3 to 8, and, in some cases, are effective at broader pH ranges, particularly when quaternized cationic monomers are used. One notable exception is cationic Mannich amines (i.e., secondary amines). Mannich amines have high charge density at pH 4 but have very low charge density ($<0.5\,g/meq$) at pH 7. Therefore, Mannich polymers are normally used in treatment systems at pH values of less than 7.

OPTIMIZING APPLICATION PROCEDURES

General Aspects

Wastewater treatment involves concentration processes in which waterborne contaminants are removed from the larger waste stream and concentrated in a smaller side stream. Proper application of wastewater treatment polymers is critical to ensure the most efficient process. Three key factors that can maximize polymer performance and minimize the amount of residual polymer lost in the effluent are:

1. Polymer selection.
2. Polymer preparation.
3. Plant system conditions, including process and operational variations.

Polymer Selection

Choice of the optimum polymer is dependent on equipment design and characteristics of the raw water to be treated. Although the nature of a waste stream is often variable, initial characterization of wastewater is essential to identify water conditions that preclude certain polymer types. Characterizing wastewater conditions according to pH, total solids, hardness, alkalinity, and chlorine will assist in the selection of the proper polymer. For instance, the effectiveness of cationic water treatment polymers is dependent on wastewater pH (Table 7.3).

Selecting the proper polymer and the optimum dosage for achieving treatment objectives can be accomplished by conducting bench-scale laboratory tests that simulate plant processes. The method chosen should simulate the plant process to which the polymer will be applied. Table 7.4 lists laboratory methods that have been developed to simulate clarification, flotation, sludge thickening, sludge dewatering, and tertiary (polishing) treatments. For example, Figure 7.3 presents results of laboratory testing to select the proper polymer for a dewatering application for the primary sludge of a paper production plant. Of the four polymers evaluated, only one (AETAC-containing polymer with 10% ionicity) achieved suitable results at a dosage between 20 and 30 mg/L.

Table 7.3. Effective pH ranges of cationic polyelectrolytes

Polymer class	Effective pH
Cationic, tertiary amine	<7
Cationic, quaternary ester	<9
Cationic, quaternary	<11

Table 7.4. Laboratory methods for simulating treatment processes[a]

Test type	Purpose/simulation
Jar tests	Clarification processes
Refiltration rate	Filtration processes
Flotation cells	Induced air or dissolved air flotation processes
Buchner funnel test	Thickening and dewatering applications
Capillary suction time test	Thickening and dewatering applications
Filter leaf test	Vacuum dewatering treatments

[a]For further information, see Betz Laboratories, Inc. (1991), Dentel et al. (1993), and Schwoyer (1981).

Polymer Preparation

Once the most appropriate polymer has been chosen, it must be applied ("madedown") properly at a suitable addition point to ensure maximum efficiency. Polymers are available in three forms: powders, liquids, and emulsions. Each of these physical states has different feeding, handling, and storage requirements, and care must be taken to follow the recommended procedures for application (Schwoyer, 1981).

Temperature, hardness, and pH of the application ("makedown") water will affect the dilution and/or the activation of water treatment polymers. Knowledge of the makedown water characteristics will allow for adjustments to be made to optimize polymer performance and minimize insuffi-

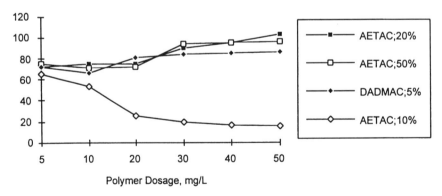

Figure 7.3. Capillary suction test (CST) method for the primary sludge of a paper production plant. (*Source*: Betz Laboratories, Inc., 1995.)

cient activation. Further, physical adjustments will not only impact performance but also affect the amount of residual polymer lost in effluent streams. For example, polymer viscosity will affect the level of mixing. Too much mixing during makedown will shear the polymer, decreasing treatment performance. Too little mixing can result in inferior performance, possibly leading to overdosing of the polymer.

Proper polymer feed points are critical to ensuring optimum polymer performance. Contact time is critical to facilitate maximum aggregation of particulates. Insufficient time for mixing with the wastewater can lead to poor performance and, in some instances, effluents containing residual polymer. Feed points thus should be chosen to maximize contact time between water treatment polymers and wastewater contaminants.

Plants System Conditions

Both process and operational variations within industrial or municipal water treatment plants can impact the treatability of a waste stream. Process variations are specific changes in the waste stream, and operational variations are overall changes that affect the operation of the plant.

Typical process variations that will influence treatability and may cause errors in dosing are:

- *Variations in suspended solid concentrations.* Optimizing polyelectrolyte dosage is a function of controlling the ratio of the amount of polymer to the weight concentration of the solids in the system. If the solids concentration suddenly varies significantly, either underdosing or overdosing can occur.
- *pH Variations.* Changes in wastewater pH can greatly affect polymer performance. If a polymer has been selected that is pH-sensitive, high pH excursions (e.g., caused by a caustic dump or process changes) can render the polymer ineffective or require higher dosages for effective performance. Changes of pH can also affect the surface charge of particulates, thereby altering the overall charge demand of the system.
- *Variations in ion concentrations.* Changes in the wastewater electrolyte concentrations can affect dissolution and uncoiling of wastewater treatment polymers, altering performance. Ion concentration will also significantly influence colloid stability and charge demand.
- *Organic variations.* As with solids, changes in the process or unexpected system upsets can change the polymer demand of the system, resulting in dosing errors.

Operational variations include the following:

- *Temperature changes due to climate.* Polymer viscosity is correlated with temperature conditions. Therefore, temperature can affect both the makedown and the mixing of wastewater treatment polymers.
- *Volume variability.* Depending upon the daily operation of the plant, the flow to the wastewater plant can fluctuate. If the polymer dosage has been optimized to a constant volume, fluctuations will result in dosing errors.
- *Compositional variability.* If the waste stream is a combination of different wastewater streams, volume ratios of the streams can vary. The polymer dosage usually is optimized to the average of ratios typically seen. Changes in the ratios of wastewater streams can lead to underdosing or overdosing, or require selection of water treatment polymers with alternative performance characteristics.

Maintaining polymer applications at optimum dosages is not only critical for achieving cost-effective treatment objectives; optimization also helps to eliminate excess quantities of residual wastewater treatment polymers that remain in solution and to bypass liquid-to-solid separation.

CHARACTERIZATION OF EXPOSURE

General Aspects

Characterizing the exposure of water treatment polymers to ecosystems requires an evaluation of polymer partitioning from the liquid to the solid state, degradation potential, degree of mobility, and fate of the residual polymers.

Degradation and Sorption Potential

Degradation of these high molecular weight polymers by biodegradation or by other ecological degradation processes (e.g., hydrolysis, volatilization, photolysis) is limited. Adsorption is the primary process that will control ecological concentrations and the environmental mobility of water treatment polymers. In spite of their initially high water solubility, water treatment polymers rapidly react with many kinds of naturally occurring substances upon reaching aquatic and soil environments (e.g., humic acids, lignins, silts, and clays). The ecological effects section in this chapter compares the toxicity of bound and soluble polymers. Proper procedures in administering polymers to treatment systems can maximize partitioning of the polymer to the sludge and minimize ecological exposure to water treatment polymers.

The efficiency of water treatment polymers moving from liquid waste streams to an aggregated solid state is often greater than 99%. Several studies were conducted to assist the U.S. Environmental Protection Agency (U.S. EPA) in ecological fate and effects assessment of low molecular weight cationic and nonionic polymers (Podoll and Irwin, 1988; Podoll et al., 1987). This research generated polymer adsorption isotherms with various sediments, and showed characteristic Langmuir adsorption isotherms with high initial slopes and sorption values that plateaued proportionally to the cationic exchange capacity or the clay content of the sediments. The strength of polymer adsorption, partitioning, and adsorption capacity varied, depending on the polymers' molecular weight, molecular structure, concentration, and electrostatic interaction potential. For example, two cationic polymers (oligomers), polyethylenimine (PEI) and polydimethylaminoethyl methacrylate (PDAM), had adsorption capacities that correlated with the cationic exchange capacity of the sediments. The adsorption capacity of nonionic polyethylene glycol (PEG) correlated with the clay content of the sediments. All three adsorbed strongly, particularly at low concentrations, to the sediments. In addition, desorption of PEI, PDAM, and PEG was less than 15% at low concentrations. Desorption of PEI and PEG only increased in the plateau region of the adsorption isotherm where there was high surface coverage of the sediments by the polymers. The high surface coverage may have prevented direct attachment of PEI and PEG to the sediment surface. In contrast, PDAM desorption was low at all surface coverages. PDAM will attach irreversibly by interacting with protonated nitrogens directly attached to cationic exchange sites.

Strong adsorption by "multisegment" attachment is expected to be more pronounced with higher molecular weight polymers. Based on studies with cationic oligomers, many segments of a polymer molecule can attach to a surface, and it is statistically improbable that all polymer segments will desorb simultaneously (Podoll et al., 1987; Podoll and Irwin, 1988). Another study has assessed the adsorption mechanisms of an anionic polyacrylamide and a monomer, acetamide, to clays (Stutzmann and Siffert, 1977). Adsorption of polyacrylamide and acetamide was irreversible. The degree of adsorption depended upon the cationic exchange capacity and the surface area of the clays.

Optimization and Monitoring

The preceding section on water treatment applications emphasized the importance of proper administration procedures for achieving optimal dosages. Simulation methods (see above) are available to aid in achieving optimal performance. Further, the monitoring of treatment processes can be used for early notification in preventing the release of residual polymers and to optimize polymer performance.

Disposal Methods

Sludge that is generated from industrial and municipal wastewater treatment systems is the primary repository for many water treatment polymers. Methods of disposal of sludge depend upon government regulations (e.g., the U.S. EPA Resource Conservation and Recovery Act), local requirements, geographical location, and sludge characteristics. Common disposal methods include incineration, land application, and landfill.

Waste management of sludge by incineration is becoming a favored practice in some areas, particularly with a reduced number of landfill sites available. Sludge incineration is typically a two-step process involving drying and complete combustion. In some locations, sludge from the biological treatment of wastewater can be used as fertilizer or for soil conditioning. When sludge is used for soil conditioning and/or agricultural purposes, analysis of the sludge is important to evaluate the potential for exposure to substances such as toxic organics, heavy metals, and inorganic nitrogen. As conventional alternatives to land application and incineration, landfills are used for disposal of industrial wastewater sludge. Impermeable liners, leachate collection, and treatment systems often are required, as well as steps to reduce leachate potential by decreasing sludge moisture content. Given the lack of mobility associated with irreversible adsorption to solids, landfill leachate will not contain ecologically significant levels of water treatment polymers.

Releases to Surface Water

The release of residual wastewater treatment polymers to surface waters can be attributed to overloading of polymers during wastewater treatment. However, ecologically significant releases of polymers from primary treatment sources (e.g., primary clarification, sludge processing, and filtration) will be rare. If overdosing occurs during primary treatment, within-plant downstream secondary and tertiary treatment should provide sufficient removal of polymers by adsorption. Polymer use in secondary and tertiary treatment units downstream in the process may have greater potential for release because there is less potential for polymer adsorption before discharge.

ECOLOGICAL EFFECTS

General Aspects

As noted above, aquatic environments are primary potential compartments of exposure to water treatment polymers; so most research with these polymers has focused on their effects in aquatic species. Aquatic effects of water treatment polymers are dependent on polymer bioavailability. Because the basic function of water treatment polymers is to adsorb waterborne contaminants to form solids, this section assesses the aquatic toxicity of both liquid and solid polymer phases.

Microorganisms

Microbial inhibition studies (Amos, 1996) have been conducted with anionic and cationic polymers to screen for potential effects on microorganisms in biological-based treatment systems. Microbial inhibition was determined by monitoring the assimilation of ^{14}C-labeled amino acids in the presence of polymer. The inoculum used in these studies was a mixed bacterial population. For seven of the nine anionic polymers tested, no inhibition was expressed at 100 mg/L as active polymer. The remaining two anionic polymers gave 25% and 60% inhibition. The cationic polymers expressed a greater inhibitory effect than the anionic polymers. At 100 mg/L, nine of the sixteen cationics tested inhibited carbon assimilation from 80% to 97%, three were moderately inhibitory (40–50%), and four demonstrated minimal inhibition (less than 17%). Inhibition of the microbial activity of a biological treatment system would primarily be of concern in the event of an overdose of polymer originating from (1) recycling of a polymer from a specific application (e.g., sludge processing) back to biological or aeration lagoons or (2) carryover from upstream treatment systems. However, the high solids and dissolved organic carbon (DOC) of typical waste streams will have a significant buffering capacity to reduce polymer effects on sludge biomass microorganisms. Although more data would improve the assessment of polymer effects on sludge microorganisms, the inhibition studies suggest that significant toxicity will not occur at typical dosages.

Algae

Under certain laboratory test conditions, moderate toxicity to algae has been reported for anionic polymers (Nabholz et al., 1993). However, results of growth inhibition (EC_{50}) studies in green algae with polycarboxylic acids actually range quite widely, from approximately 1 mg/L to over 100 mg/L. Toxicity in algae may result from sequestration of essential elements by the polymers because toxicity to algae is substantially mitigated when the hardness of the algal growth medium is increased with the addition of calcium and magnesium ions. It has been indicated that an increase in the hardness of a standard guideline algal growth medium (i.e., 15 mg/L calcium carbonate) to the average hardness of natural water (i.e., 120–150 mg/L calcium carbonate) would present realistic algal toxicity values for anionic polymers (Hamilton et al., 1994).

Based on aquatic toxicity information presented to the U.S. Environmental Protection Agency, cationic water treatment polymers are approximately six times more toxic to green algae than to fish when acute lethality data are compared (Nabholz et al., 1993). Further, as described later in this chapter, the toxicity in fish of cationic polymers is highly mitigated in the presence of natural levels of dissolved organic carbon.

Daphnids and Fish — Overview

Table 7.5 presents LC_{50} values in fish and daphnids for a wide range of water treatment polymers — cationic, anionic, and nonionic. The polymers are grouped according to the polymer classification in Table 7.2 as well as their chemistry, molecular weight, percent ionicity, and physical state. It should be noted that LC_{50} values in Table 7.5 are based on percent polymer active for each product. The results were obtained from 48- to 96-hr exposures in standard static, static-renewal, or flow-through aquatic toxicity protocols.

As emulsions, the toxicity of anionic and nonionic polymers was comparable to the toxicity of cationic polymers. Anionic and nonionic polymers in the powdered form generally had the lowest toxicity. Other useful comparisons using geometric means of the LC_{50} values of cationic water treatment polymers and all polymers combined are presented in Table 7.6. The geometric mean LC_{50} value of cationic polymers in fish was comparable to the geometric mean LC_{50} value of cationic polymers in daphnids. However, in comparing geometric mean LC_{50} values for all polymers combined, the fish LC_{50} (2.06 mg/L) was three times greater than the daphnid LC_{50} (0.7 mg/L). This analysis shows that water treatment polymers should be assessed for fluctuation in cross-species sensitivity according to charge category, not by all polymers combined.

Daphnids

LC_{50} values for three daphnid species (*Daphnia magna, Daphnia pulex,* and *Ceriodaphnia dubia*) ranged from 0.03 mg/L to 470 mg/L for all polymers in Table 7.5. Approximately 70% of the LC_{50} values were less than 1.0 mg/L, and 10% of the LC_{50} values ranged between 1 mg/L and 10 mg/L. The remaining LC_{50} values were greater than 10 mg/L. *Ceriodaphnia* tend to be the most sensitive daphnid species, with LC_{50} values that were usually one to two orders of magnitude lower than those for *D. magna.*

Cationic polymers that were least toxic to daphnids were Mannich polymers, melamine-formaldehyde (MF) polymers, and cellulosic-based polymers. LC_{50} values for Mannich polymers to *D. magna* and *D. pulex* ranged from 41.2 mg/L to 114 mg/L. As noted above, Mannich water treatment polymers have high ionicity, ranging from 70% to 100% at pH 4, but have very low ionicity at pH 7. Because aquatic toxicity tests are usually conducted at neutral pH, the potential for acute effects from ionicity, if any, is considerably reduced during toxicity testing. Likewise, MF and cellulosic-based polymers have low ionicity at pH 7. In addition, MF polymers are self-precipitating, thereby reducing their bioavailability during toxicity testing.

Nonlinear dose–response curves and wide confidence intervals generated from water treatment polymer toxicity tests with daphnids have been reported by several investigators. For example, 10 to 12 test concentrations were required with *D. magna* to define LC_{50} values (Cary et al., 1987). In approximately 30% of the tests conducted with *D. magna* in which LC_{50} values could be estimated, lethality did not follow a typical dose–response curve. Figure 7.4 presents some abnormal dose–responses following 48-hr exposures to several cationic and anionic polymers. The polymers consistently produced either a bell-shaped curve or a flattening of the curve at midlevel test concentrations for lethality.

It is important to note that water treatment polymer studies have reported that expired daphnids are found clumped together at the bottom of the test vessels. The polymers can act as "flocculants" of the test species by entrapping and clumping the daphnids together. Another study (Hall and Mirenda, 1991) also reported the physical toxicity of polymers to *D. pulex* and found that LC_{50} values of the polymers were difficult to estimate, giving wide confidence intervals.

Biomarking of polymer exposure has been performed recently with an electrochromic dye that detects changes in membrane transport kinetics

Table 7.5. LC$_{50}$ values (mg/L) based on percent polymers for cationic, anionic, and nonionic polymers in daphnids and fish

Polymer[c]	Type[d]	Chemistry[e]	MW[f]	% Ionicity	Daphnid species[a]			Fish species[b]			Ref.[h]
					DM[g]	DP[g]	CD[g]	FM[g]	RT[g]	BG[g]	
Cationic											
C1S	IIA1a	METAC	H	100	<1.0(s)			0.3(s)			1
C2P	IIA1a	METAC	H	81				<1.0(s)			1
C3E	IIA1a	METAC	H	75		0.1(s)		0.4(s)			2
C4E	IIA1a	METAC	H	45		0.12(s)		0.61(s)			2
C5E	IIA1a	METAC	H	45		0.06(s)		1.4(s)			2
C6E	IIA1a	METAC	H	25		0.19(s)		1.4(s)			2
C7E	IIA1a	METAC	H	10		0.22(s)		4.7(s)			2
C8P	IIA1a	AETAC	H	7				3.3(s)			1
C9E	IIA1a	AETAC	H	80	0.16(s)						1
C10E	IIA1a	AETAC	H	52	0.16(s)			1.1(s)			1
C11E	IIA1a	AETAC	H	50	0.08(r)			0.45(r)	0.4(s)	1.1(s)	1
C12E	IIA1a	AETAC	H	45		0.57(s)		0.8(s)			2
C13E	IIA1a	AETAC	H	45		0.98(s)		1.17(s)			2
C14E	IIA1a	AETAC	H	45		0.32(s)		2.5(s)			2
C15E	IIA1a	AETAC	H	45		0.19(s)		1.3(s)			2
C16E	IIA1a	AETAC	H	40	1.0(s)			0.9(s)	0.4(s)	1.0(s)	1
C17E	IIA1a	AETAC	H	40	0.04(s); 1.0(r)		0.03(r)	1.1(s); 2.1(r)			1
C18	IIA1a	AETAC	H	39					0.66(s); 0.38(f)		3
C19E	IIA1a	AETAC	H	35		0.21(s)		1.05(s)			2
C20E	IIA1a	AETAC	H	25		0.2(s)		1.45(s)			2
C21E	IIA1a	AETAC	H	20	1.2(s)			3.1(s)			1
C22E	IIA1a	AETAC	H	20	17(s); >19.5(r)		<0.04(r)	3.6(s); 7.1(r)			1
C23E	IIA1a	AETAC	H	15	0.4(r)			0.6(r)			1
C24E	IIA1a	AETAC	H	10	0.6(s)			2.9(s)			1
C25E	IIA1a	AETAC	H	10	1.4(r)			3.4(r)			1

Sample	Type	Chemistry	Charge	%							N
C26E	IIA1a	AETAC	H	10		0.15(s)	4.49;3.29(s)				2
C27	IIA1a	AETAC	H	10		0.06(s)	13.5(s)		1.7(s)		3
C28E	IIA1a	AETAC	H	6			0.8(r)		0.3(s)		2
C29E	IIA1a	AETAC	H	7	10.1(r)		1.4(s);8.2(r)	0.06(r)			1
C30E	IIA1a	AETAC	H	5	0.9(s);6.1(r)		11.6(s)				1
C31E	IIA1a	AETAC	H	2	0.17(s)		0.6(r)				1
C32E	IIA1a	AETAC	H	1	0.17(s)		1.2(s)				1
C33E	IIA1a	MAPTAC	H	<5	2.3(s)		0.2(s)				1
C34S	IA1	EPI/DMA	L	100	0.07(s)		0.23(r)			0.18(s)	1
C35S	IA1	EPI/DMA	H	100	0.31(s)		0.86(s)		0.2(s)	0.26(s)	1
C36S	IA1	EPI/DMA	M	100		0.26(s)	0.68(s)				2
C37S	IA1	EPI/DMA	M	100		0.16(s)					2
C38S	IA1	EPI/DMA	M	100	0.08;0.17(s)		0.25;0.30(s)			0.18(s)	4
C39S	IA1	EPI/DMA	L	—					0.59(s);0.04(f)		3
C40S	IA1	EPI/DMA	L	—					0.27(s);0.09(f)		3
C41S	IA1	EPI/DMA	M	—					0.78(s);0.15(f)		3
C42E	IIA2	DADMAC	H	2	4.5(s)		5.4(s)				1
C43S	IA2	DADMAC	M	100	0.25(s)	0.77(s)	0.2(s)		0.09(s)		1
C44S	IA2	DADMAC	L	100		2(s)	0.74(s)				2
C45S	IA2	DADMAC	M	100			0.88(s)				2
C46S	IA2	DADMAC	L	100	0.2(s)		0.46(s)				4
C47S	IB1	Polyamine	M	—	2.2(r)		0.3(r)		0.7(s)	1.5(s)	1
C48S	IB1	Polyamine	M	—	0.3(r)		0.1(r)		0.4(s)	0.5(s)	1
C49S	IB1	Polyamine	L	100	0.3(s)						1
C50S	IIA2a	Mannich	H	70		46.2(s)	1.19(s)				2
C51S	IIA2a	Mannich	H	70		70.1(s)	1.36(s)				2
C52S	IIA2a	Mannich	H	70		45.9(s)	1.04(s)				2
C53S	IIA2a	Mannich	H	70		41.3(s)	1.48(s)				2
C54S	IIA2	Mannich	H	70		51.7(s)	3.29(s)				2
C55S	IIA2a	Mannich	H	100	114(r)		0.7(r)			0.9(s)	1
C56S	IC1	MF	L	100			8.0(s)				1

Table 7.5. Continued

Polymer[c]	Type[d]	Chemistry[e]	MW[f]	% Ionicity	Daphnid species[a]			Fish species[b]			Ref.[h]
					DM[g]	DP[g]	CD[g]	FM[g]	RT[g]	BG[g]	
Cationic (Continued											
C57S	IC1	MF	L	100	16(s)			>170(s)			1
C58S	IC1	MF	L	75		12.1(s)					2
C59	IC2	Pyridine	H	77	0.08(s)			0.17(s)		0.06; 0.2(s)	4
C60S	IC3	Cellulosic	L	100	32(r)			47.5(r)			1
C61S	IC3	Cellulosic	L	100	92(r)			0.6(r)			1
C62	IC3	Polyalkamine	L	80	0.21(s)		0.5(r)	0.16; 0.18(s)		0.32; 1.0(s)	4
Anionics											
A1P	IIB1a	AA,AM	H	−39	470(s)						
A2P	IIB1a	AA,AM	H	−31				810(s)	>100(s)	>300(s)	1
A3P	IIB1a	AA,AM	H	−22					>100(s)	>300(s)	1
A4P	IIB1a	AA,AM	H	−12					>100(s)	>300(s)	1
A5E	IIB1a	AA,AM	H	−100	0.7(r)			2.2(r)			1
A6E	IIB1a	AA,AM	H	−100	0.6(s)				0.3(s)		1
A7E	IIB1a	AA,AM	H	−33	0.4(s)						1
A8E	IIB1a	AA,AM	H	−30		0.41; 0.62(s)		85.1(s)			2

A9E	IIB1a	AA,AM	H	−30		0.06(s)	20.9(s)	2
A10E	IIB1a	AA,AM	H	−30		0.39(s)	28.4(s)	2
A11E	IIB1a	AA,AM	H	−30		0.66(s)	36(s)	2
A12E	IIB1a	AA,AM	H	−26	0.4(r)		0.8(r)	1
A13E	IIB1a	AA,AM	H	−20		0.11(s)	40.4(s)	2
A14E	IIB1a	AA,AM	H	−20	0.1(r)		3.5(r)	1
A15E	IIB1a	AA,AM	H	−8		0.09(s)	37.2(s)	2
A16E	IIB1a	AA,AM	H	−5	0.11(s)			1
A17P	IIB1b	AMPS,AM	H	−10			1340(s)	1
Nonionics								
N1P	IIC1	PA	H	0	350(s)		465(s)	1
N2E	IIC1	PA	H	0	0.04(s)		41(s)	1
N3E	IIC1	PA	H	0		0.08(s)	63.6(s)	2
N4E	IIC1	PA	H	−4		0.15(s)	63.6(s)	2

[a] DM, *Daphnia magna*; DP, *Daphnia pulex*; CD, *Ceriodaphnia dubia*.

[b] FM, *Pimephalus promelas*, fathead minnow; RT, *oncorhynchus mykiss*, rainbow trout; BG, *Lepomis macrochirus*, bluegill sunfish.

[c] C, cationic; A, anionic; N, nonionic; E, emulsion; P, powder; S, solution.

[d] Refer to Table 7.2.

[e] Refer to Table 7.2.

[f] MW, molecular weight (L < 100,000: M = 100,000 to 1,000,000: H > 1,000,000).

[g] s, static; r, static-renewal; f, flow-through.

[h] 1 (Betz Laboratories, Inc., 1995), 2 (Hall and Mirenda, 1991), 3 (Goodrich et al., 1991), 4 (Cary et al., 1987).

Table 7.6. Geometric means (mg/L) of polymer LC_{50} values [a,b]

	All polymers	Cationic polymers
All fish species	2.06(98)	0.82(74)
P. promelas	2.6(69)	1.2(54)
O. mykiss	0.79(16)	0.4(12)
L. macrochirus	1.86(13)	0.25(10)
All daphnid species	0.7(74)	0.72(57)
D. magna	0.89(40)	0.82(31)
D. pulex	0.59(31)	0.9(23)
C. dubia	0.08(4)	0.08(4)

[a]Means (mg/L) derived from Table 7.5; all " < " and " > " values in Table 7.5 were used as absolute values, and only lowest LC_{50} values in Table 7.5 were used in cases with replicate data.
[b]( ) = number of LC_{50} values used in calculation.

(Fort et al., 1966). *Daphnia magna* were exposed to sublethal levels of a polyacrylamide in the presence of dye, and very little of the fluorescence was internalized. The most intense fluorescence was located in epithelial cells on the surfaces (integument) of the exposed organisms. This evidence of polymer-related effects on integument membrane transport supports the hypothesis that cationic polymer toxicity in daphnids results from surface membrane effects (e.g., flocculation) that alter transport through the integument and/or inhibit appendage movement that would normally provide adequate nutrient uptake.

The toxicity of cationic quaternary amine flocculant polymers to *C. dubia* may be increased when they are used with inorganic coagulant aids

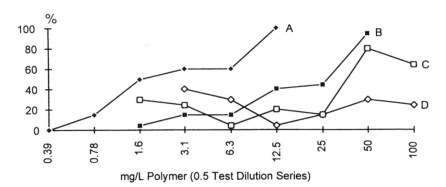

Figure 7.4. Nonlinear *D. magna* mortality dose–responses, where A is polymer C17E, B is polymer C30E (see Table 7.5 for structure of C17E and C30E), C is cationic AETAC-containing polymer, and D is anionic polymer. (*Source*: Betz Laboratories, Inc., 1995.)

(Fort and Stover, 1995). However, the lower LC_{50} values being reported for *C. dubia* may suggest that this smaller daphnid is more susceptible to physical toxicity (e.g., flocculation) by water treatment polymers than larger daphnid species with their greater surface area.

Fish

Cationic polymers are more toxic in acute testing to the freshwater fish species (i.e., *Pimephales promelas*, *Oncorhynchus mykiss*, and *Lepomis macrochirus*) than anionic and nonionic polymers (Table 7.5). With the exception of four LC_{50} values, the anionic and nonionic polymers had LC_{50} values above 20 mg/L, and several polymers had LC_{50} values above 100 mg/L. Emulsion anionic and nonionic polymers expressed lower LC_{50} values than powder forms. Cationic solution polymers (EPI/DMA-containing, DAD-MAC-containing, and polyamines) with 100% ionicity tended to be more toxic than cationic emulsion polymers (AETAC and METAC).

It is useful to note that Mannich polymers with their low charge densities at neutral pH values were more toxic to fish than to daphnids (see above). This suggests that tissue surface bioavailability of Mannich polymers to fish could be a function of hydrophobic interactions, not electrostatic attraction.

Whether by hydrophobic or electrostatic mechanisms, the affinity of anionic, cationic, or nonionic water treatment polymers for gill epithelial tissue is considered primarily responsible for inducing toxicity in fish. Physical damage to fish gill tissue by cationic polymers has been reported (Biesinger and Stokes, 1986). Examination of fathead minnow gill filaments by microscopy revealed that increased cationic polymer dosages and duration caused gradual deterioration and distortion of the gill structures to the extent that those structures were not definable. Another study (Kosteretz et al., 1994) exposed fathead minnows to ^{14}C-labeled epichlorohydrin dimethylamine to identify the accumulation of the polymer distributed in gills, viscera, muscle, and skin. Only the gills were identified as having ^{14}C levels above background levels. These findings indicate that cationic polymers interact with gill tissue, disrupting gill structure and function.

Other Aquatic Organisms

Aquatic toxicity data are limited for benthic (i.e., sediment) organisms, marine or estuarine species, and aquatic insects. However, an important study showed that cationic polymers in soluble form were less toxic to midges (*Paratanytarus*) and gammarids (*Gammarus pseudolimnaeus*) than to *D. magna* and fathead minnows (Biesinger and Stokes, 1986). The 96-hr LC_{50} for gammarids with seven of the thirteen cationic polymers ranged from 8.1 mg/L to 85.2 mg/L (mg/L was based upon formulation concentration, not active polymer), with the remaining six polymers having LC_{50} values above 100 mg/L. Of the eight cationic polymers tested with midges, five polymers had LC_{50} values above 100 mg/L, with the remaining polymers ranging from less than 6.25 mg/L to 50 mg/L. Additional studies would be useful to further assess the effects of bound polymer on sediment organisms, but the toxicity (and bioavailability) of bound water treatment polymers will be less than the toxicity of free soluble polymers (see below).

Chronic Toxicity

Tables 7.7 to 7.10 provide data from long-term sublethal exposures, partial life-cycle tests, and early life-stage survival and growth studies of representative water treatment polymers. Flow-through 28-day rainbow trout studies assessed both survival and sublethal effects (growth) from exposures to two EPI/DMA-containing polymers and an AETAC-containing polymer (Goodrich et al., 1991). Virtually all lethality occurred within the first 4 days of exposure. Low acute to chronic LC_{50} ratios of 1.0 to 1.3 (Table 7.7) further indicated that the degree of toxicity was essentially the result of abrupt acute effects.

Trout that survived the 28-day polymer exposures were measured for wet body weight and total length. Table 7.8 provides no effect concentrations (NOEC) for growth. Growth impairment did not occur at all doses

Table 7.7.   Flow-through polymer exposures to rainbow trout for 4-, 7-, and 28-day test durations

	LC_{50} (mg/L)		
Exposure period	**C18[a]**	**C39[a]**	**C41[a]**
4-day	0.38	0.04	0.16
7-day	0.36	0.04	0.15
28-day	0.30	0.04	0.14
4 day/28 day ratio	1.3	1.0	1.1

[a]See Table 7.5 for polymer identification.

Source: Goodrich et al., 1991.

Table 7.8. NOEC values (mg/L) from 28-day polymer exposures to trout growth: Wet weight and length

	C18[a]	C39[a]	C41[a]
Weight	0.108	>0.049[b]	>0.098[b]
Length	>0.215[b]	>0.049[b]	>0.098[b]

[a]See Table 7.5 for polymer identification.

[b]NOEC values are greater than highest test concentration.

Source: Goodrich et al., 1991.

tested, except for fish weight with 0.108 mg/L of polymer C18 (cationic AETAC-containing polymer with 39% ionicity).

Low acute to chronic ratios have been found with three cationic polymers in *D. magna* 21-day chronic studies (Biesinger et al., 1976). For these studies, the 21-day threshold level (TL_{50}) values for survival (Table 7.9) were higher by an order of magnitude than the 48-hr TL_{50} values. During the 21-day exposures with daphnids, test solutions were renewed several times along with new organic matter as a food source. The organic matter present probably reduced polymer bioavailability by adsorbing with the polymer. No significant impairment to reproduction was found at 0.1 mg/L of cationic polymer 1 and at 1.0 mg/L of cationic polymers 2 and 3. Doubling of the dose for each polymer, however, resulted in reproductive impairment.

Low acute to chronic ratios for a cationic polyacrylamide emulsion polymer have been reported with *Ceriodaphnia dubia* and *Pimephales promelas* (fathead minnow); the ratios were 2.9 and 2.4, respectively (God-

Table 7.9. *Daphnia magna* acute and chronic test results (mg/L) from cationic polymer exposures

	Polymer 1	Polymer 2	Polymer 3
Survival:			
4-day TL_{50}	0.34	0.65	—
21-day TL_{50}	1.1	2.85	1.85
Reproduction:			
No impairment	0.1	1.0	1.0
Significant impairment	0.2	2.0	2.0

Source: Biesinger et al., 1976.

Table 7.10. Acute to chronic toxicity ratios of cationic polyacrylamide wih *Ceriodaphnia dubia* and fathead minnow

Test organism	LC_{50} (mg/L)	NOEC (mg/L)	Acute/chronic ratio
P. promelas	7.5	3.1	2.4
C. dubia	0.5	0.17	2.9

Source: Godwin-Saad et al., 1994.

win-Saad et al., 1994). Table 7.10 summarizes acute and 7-day chronic bioassays for *Ceriodaphnia* reproduction and fathead minnow larvae growth.

Chronic testing protocols can reduce the bioavailability of water treatment polymers by allowing polymer adsorption to organic food sources. Standard studies do not directly determine if adsorbed polymer is ingested or merely becomes unavailable by flocculating and/or settling, but loss of food by polymer adsorption can interfere with testing systemic chronic toxicity. However, in spite of potential study artifacts, some useful trends are evident. Abrupt lethality and low acute to chronic ratios indicate that the toxicity of water treatment polymers is best correlated with acute effects. The ability of water treatment polymers to bioaccumulate is limited by high molecular weight and rapid partitioning to organics.

Toxicity of Bound Polymers

Polymers that have been partitioned to various substrates (i.e., clays, silts, humic acid, biosolids, and sludge) have demonstrated significant reduction of toxicity to aquatic organisms. Reduction of acute toxicity of cationic polymers by adsorption to suspended solids (SS) and dissolved organic carbon (DOC) is shown in Table 7.11 (Cary et al., 1987). For each of the above tests, a range of polymer concentrations was used in the presence of 50 mg/L as SS or 10 mg/L as DOC. Concentrations of SS and DOC used in the study were considered to be conservative (low) estimates of SS and DOC levels found in natural environments. The studies were not designated to define organic substrate concentrations that would completely eliminate acute toxicity, but rather demonstrated that toxicity would be limited by a variety of natural organic substrates.

Consistent with the results presented in Table 7.11, Table 7.12 presents LC_{50} values to *D. pulex* and *P. promelas* for polymer C4 (cationic METAC-containing polymer with 45% ionicity), which progressively increase with the addition of humic acid in the dilution water (Hall and Mirenda, 1991). Addition of up to 60 mg/L humic acid (approx. 9.6 mg/L organic carbon)

Table 7.11. Reduction of acute toxicity to *D. magna* and fathead minnows by suspended solids and dissolved organics

Substrate	Polymer C38[a]		Polymer C46[a]		Polymer C62[a]	
	Daphnids	Fatheads	Daphnids	Fatheads	Daphnids	Fatheads
Bentonite[b]	75 ×	26 ×	36 ×	14 ×	96 ×	46 ×
Illite[b]	6.9 ×	3.8 ×	6.0 ×	1.2 ×	4.8 ×	6.9 ×
Kaolin[b]	11 ×	2.6 ×	5.5 ×	0.9 ×	4.3 ×	2.6 ×
Tannic acid[c]	100 ×	26 ×	59 ×	14 ×	83 ×	29 ×
Lignin[c]	50 ×	14 ×	>77 ×	8 ×	137 ×	24 ×
Humic acid[c]	63 ×	16 ×	37 ×	14 ×	50 ×	40 ×
Lignosite[c]	59 ×	15 ×	39 ×	8 ×	28 ×	18 ×
Fulvic acid[c]	48 ×	15 ×	11 ×	9 ×	70 ×	14 ×

[a]See Table 7.5 for polymer identification and LC_{50} values in standard laboratory water.
[b]Test conducted in presence of 50 mg/L of substrate.
[c]Test conducted in presence of 10 mg/L of substrate.
Source: Cary et al., 1987.

reduced the toxicity of the cationic polymer by almost two orders of magnitude for both species. Therefore, as indicated by Hall and Mirenda, polymer toxicity tests conducted using dilution water free of organic carbon will likely overestimate the toxicity of polymers in wastewaters. Other studies have also shown that the addition of various substrates will reduce

Table 7.12. Reduction of acute toxicity for polymer C4[a] with varying concentrations of humic acid in dilution water

	LC_{50} (95% confidence interval), mg/L	
Humic acid (mg/L)	*D. pulex*	*P. promelas*
0	0.26 (0–1.85)	0.81 (0.37–1.1)
2	0.45 (0.19–1.1)	1.1 (0.7–1.8)
10	1.2 (0.19–1.85)	6.0 (4.4–8.9)[b]
20	4.1 (1.1–5.5)	12.5 (9.2–18.5)[b]
40	7.2 (3.7–29.6)[b]	48.5 (44.4–59.2)[b]
60	13.4 (11.1–22.2)[b]	64.0 (59.2–74.0)[b]

[a]See Table 7.5 for polymer identification.
[b]Denotes LC_{50} value significantly different from LC_{50} value that was generated from humic acid–free test solutions.
Source: Hall and Mirenda, 1991.

or eliminate polymer toxicity (Biesinger and Stokes, 1986; Biesinger et al., 1976). Goodrich et al. (1991) demonstrated that 5 mg/L of humic acid reduced the toxicity of cationic polymers to fingerling rainbow trout 7- to 16-fold.

Water Treatment System Effluents

The maintenance of optimal polymer dosages to achieve appropriate treatment objectives is critical in managing effluent toxicity. Proper administration of the polymer application and proper selection of the polymer, as previously discussed, are necessary to maintain the optimal polymer treatment.

A study that included simulation of clarification of synthetic river water with cationic polymers has assessed acute toxicity from underdose to overdose conditions (Devore and Lyons, 1986). Figures 7.5 and 7.6 define the optimal dosages for clarifying synthetic river water with an EPI/DMA-containing polymer (C34; see Table 7.5 for more information) and a DADMAC-containing polymer (C43; see Table 7.5 for more information). Clarification was measured in terms of turbidity according to nephelometric

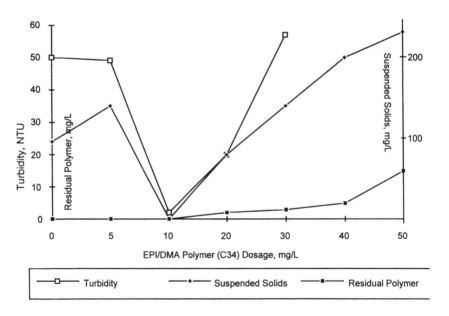

Figure 7.5. Clarification of synthetic river water with cationic polymer (C34) applications at underdose, optimal, and overdose levels. (*Source*: Devore and Lyons, 1986.)

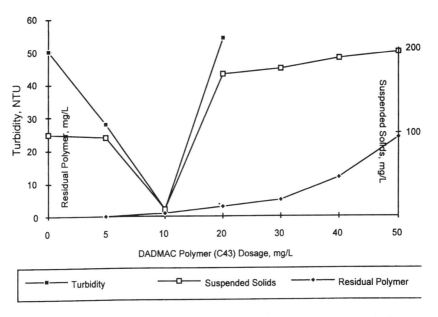

Figure 7.6. Clarification of synthetic river water with cationic polymer (C43) applications at underdose, optimal, and overdose levels. (*Source*: Devore and Lyons, 1986.)

turbidity units (NTU). The optimal treatment dosage for both polymers was 10 mg/L. Overdose treatments ranging from 20 mg/L to 40 mg/L resulted in resuspension of SS, causing turbidity to increase sharply and levels of residual polymer to become detectable.

In testing with *D. magna* and fathead minnows, no treatment-related lethality was found at 5 mg/L underdose and at 10 mg/L (Tables 7.13 and 7.14), even though 5 mg/L and 10 mg/L were one to two orders of magnitude above LC_{50} values of the same polymers in standard laboratory testing. When the EPI/DMA-containing polymer reached 30 mg/L and 40 mg/L, there was 100% mortality. The DADMAC-containing polymer caused a somewhat sharper dose–response in fathead minnows, with 20 mg/L giving 100% lethality. No mortality to *D. magna* was found with the DAD-MAC-containing polymer until the 40 mg/L overdose treatment. When combined with the wastewater treatment performance data (Figures 7.5 and 7.6), the results showed that not only does an overdose of water treatment polymer reduce performance (i.e., give increased SS levels), but the overdose will substantially increase aquatic toxicity.

Devore and Lyons (1996) have also evaluated the aquatic effects of wastewater treated with polymers. The study involved clarification of a raw

Table 7.13. Mortality responses from exposures to polymer C34[a]

	% Cumulative mortality: Fathead Minnow			
Dosage (mg/L)	1 hr	24 hr	48 hr	96 hr
5	0	0	0	0
10	0	0	0	0
20	0	0	0	0
30	5	95	100	—
40	70	100	—	—
	% Cumulative Mortality: Daphnia magna			
5	0	0	0	
10	0	0	0	
20	0	0	0	
40	5	100	—	

[a]See Table 7.5 for polymer identification and LC_{50} values in standard laboratory water.
Source: Devore and Lyons, 1986.

liquor wastewater of a food processing plant with an EPI/DMA-containing polymer (C35; see Table 7.5 for more information). Results from a mixed liquor jar test to simulate treatment of the plant's wastewater are shown in Figure 7.7. The optimal dosage reduced the suspended solids in the untreated wastewater from 3,300 mg/L to 20 mg/L. The optimal dose for this polymer was 550 mg/L. When the raw liquor received overdose treatments,

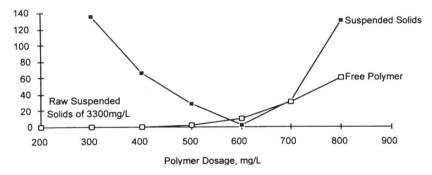

Figure 7.7. Clarification of food plant wastewater and residual polymer available at various treatment dosages. (*Source*: Devore and Lyons, 1986.)

Table 7.14. Mortality responses from exposures to polymer C43[a]

Dosage (mg/L)	% Cumulative mortality: Fathead Minnow			
	1 hr	24 hr	48 hr	96 hr
5	0	0	0	0
10	0	0	0	0
20	15	55	75	80
40	40	100	—	—

	% Cumulative Mortality: Daphnia magna		
5	0	0	0
10	0	0	0
20	0	0	0
40	10	75	100

[a]See Table 7.5 for polymer identification and LC_{50} values in standard laboratory water.
Source: Devore and Lyons, 1986.

restabilization of waste solids *prevented* clarification, and residual polymer was detected in the jar test supernatant.

For the bioassay portion of the study, it was necessary to evaluate the toxicity of the raw wastewater with and without polymer added. The raw liquor without polymer caused a 100% lethality in fathead minnows within a 24-hr exposure period (Figure 7.8). However, the toxicity of the raw

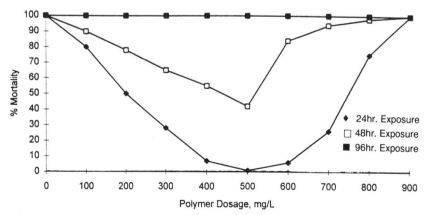

Figure 7.8. Toxicity of food plant's mixed liquor wastewater treated with polymer to fathead minnows. (*Source*: Devore and Lyons, 1986.)

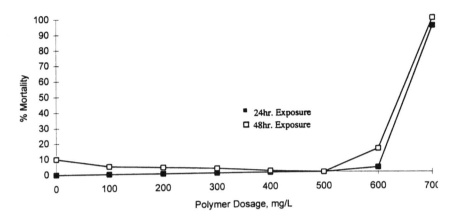

Figure 7.9. Toxicity of food plant's mixed liquor wastewater treated with polymer to *D. magna.* (*Source:* Devore and Lyons, 1986.)

wastewater was mitigated in an acute time frame with polymer, according to dose; that is, during the 24-hr exposure period there was no lethality with the optimal dosage of 550 mg/L. With additional exposure time, mortality increased to 40% at 48 hr and to 100% at 96 hr. Toxicity increased at a faster rate with overdoses, possibly as a result of restabilization of the liquor solids and availability of residual polymer. The same untreated raw wastewater from the food processing plant was not nearly as toxic to *D. magna,* resulting in 10% lethality (Figure 7.9). With polymer, the lethality was reduced to zero at the 550 mg/L optimal dosage at 48 hr. The toxicity increased sharply with overdose treatments.

Safe use with an optimal dose is possible. Water treatment polymers have been used for treatment of water supplies to fish hatcheries. For example, one study (Olson et al., 1973) evaluated cationic and anionic water treatment polymers over a two-year period. The polymers were used to reduce the turbidity of the water supply of a trout hatchery following storm events. Trout that were reared in untreated water containing high turbidity were in poor condition with reduced body weight. However, trout that were reared in treated water were in good condition, showing no apparent adverse effects on their growth.

CONCLUSIONS

Realistic risk characterization and management of water treatment polymers must consider the type of polymer application, procedures that could be responsible for releases to surface waters, the polymer fate in natural aquatic environments, and differences in bioavailability and toxicity of soluble versus solid bound polymers.

Achieving adverse ecological concentrations in aquatic environments is unlikely for upstream processes such as primary treatment. Any excess soluble polymer would be partitioned to solids in downstream processes within the same treatment system. Secondary or tertiary applications are considered more problematic, but only if water treatment polymers are being improperly applied. Appropriate measures to minimize the loss of residual water treatment polymers to surface waters include ensuring that the proper polymer is being used, that optimal administration procedures are being followed, and that the system is being adequately monitored for variability in operations and processes.

Aquatic effects, if any, from the release of free residual polymers will be a consequence of acute effects within the immediate vicinity (e.g., mixing zone) of the discharge. The high affinity of water treatment polymers to partition to natural substrates (e.g., suspended solids, humic acids, lignins, silt, and sediments) will rapidly diminish both free residual polymer concentrations and effluent toxicity within the mixing zone. The above studies have shown that (1) aquatic toxicity will be sufficiently mitigated when water treatment polymers are adsorbed to solids, and (2) aquatic toxicity is best attributed to physical effects (e.g., flocculation of daphnids during testing or gill membrane interactions in fish), not systemic effects by absorption. With their high molecular weights and rapid partitioning properties preventing bioaccumulation, water treatment polymers have a limited ability to cause chronic aquatic toxicity. Irreversible adsorption to naturally occurring substrates in water, sludges, and soils will limit their long-term mobility.

The environmental benefits (e.g., pollution control and water reuse) of water treatment polymers considerably outweigh their ecological risks. Proper application procedures and monitoring of water treatment polymers will optimize polymer performance and provide reasonable mitigation measures to minimize ecological effects.

REFERENCES

Amos, D. A. 1996. *Environmental Fate and Effects Program.* Woodlands, TX: Betz Laboratories, Inc.

Betz Laboratories, Inc. 1991. *Betz Handbook of Industrial Water Conditioning*, 9th Ed. Trevose, PA: Betz Laboratories, Inc.

Betz Laboratories, Inc. 1995. *Environmental Fate and Effects Program.* Trevose, PA: Betz Laboratories, Inc.

Biesinger, K. E. and G. N. Stokes. 1988. Effects of synthetic polyelectrolytes on selected aquatic organisms. *Journal of the Water Pollution and Control Federation* 58(3):207–13.

Biesinger, K. E., A. E. Lemke, W. E. Smith, and R. M. Tyo. 1976. Comparative toxicity of polyelectrolytes to selected aquatic animals. *Journal of the Water Pollution and Control Federation* 48:183–87.

Cary, G. A., J. A. McMahon, and W. J. Kuc. 1987. The effect of suspended solids and naturally occurring dissolved organics in reducing the acute toxicities of cationic polyelectrolytes to aquatic organisms. *Environmental Toxicology and Chemistry* 6:469–74.

Dentel, S. K., M. M. Abu-Orf, and N. J. Griskowitz. 1993. *Guidance Manual for Polymer Selection in Waste Water Treatment Plants.* Alexandria, VA: Water Environmental Research Foundation.

Devore, D. I. and L. A. Lyons. 1986. Toxicity of cationic polyelectrolyte in water treatment system effluents, in *Environmental Quality and Ecosystem Stability*, Vol. III A/B, pp. 647–58. Ramat-Gan, Israel: Bar-Ilan University Press.

Fort, D. J., and E. L. Stover. 1995. Impact of toxicities and potential interaction of flocculants and coagulant aids on whole effluent toxicity testing. *Water Environment Research* 67:921–25.

Fort, D. J., E. L. Stover, S. L. Burks, R. A. Atherton, and J. T. Blankemeyer. 1996. Utilizing biomarker techniques: Cellular membrane potential as a biomarker of subchronic toxicity, in *Environmental Toxicology and Risk Assessment: Biomarkers and Risk Assessment*, Vol. 5, ed. D. Bengston and D. Henshel. Philadelphia, PA: American Society of Testing and Materials.

Godwin-Saad, E., W. S. Hall, and D. Hughes. 1994. An evaluation of the acute and chronic toxicity of a waste water treatment polymer to aquatic and terrestrial organisms, in *Proceedings of the Water, Environment Federation 67th Annual Conference and Exposition, Surface Water Quality and Ecology*, Vol. 4, p. 249.

Goodrich, M. S., L. H. Dulak, M. A. Friedman, and J. J. Lech. 1991. Acute and long-term toxicity of water-soluble cationic polymers to rainbow trout and the modification of toxicity by humic acid. *Environmental Toxicology and Chemistry* 10:509–15.

Hall, W. S. and R. J. Mirenda. 1991. Acute toxicity of waste water treatment polymers to *Daphnia pulex* and the fathead minnow and the effects of humic acid on polymer toxicity. *Research Journal of the Water Pollution Control Federation* 63:895–99.

Hamilton, J. D., K. H. Reinert, and M. B. Freeman. 1994. Aquatic risk assessment of polymers. *Environmental Science and Technology* 28:187–92.

Kosteretz, K. G., M. A. Friedman, and J. J. Lech. 1994. Mode of toxicity of a cationic polymer to rainbow trout (*Oncorhynchus mykiss*), paper presented at the Society of Environmental Toxicology and Chemistry, 16th Annual Meeting, Denver, CO.

Nabholz, J. V., P. Miller, and M. Zeeman. 1993. Environmental risk assessment of new chemicals under the Toxic Substances Control Act (TSCA) Section Five, in *Environmental Toxicology and Risk Assessment*, ASTM STP 1179, ed. W. G. Landis, J. S. Hughes, and M. A. Lewis. Philadelphia, PA: American Society for Testing and Materials.

Olson, W. H., D. L. Chase, and J. N. Hanson. 1973. Preliminary studies using synthetic polymers to reduce turbidity in a hatchery water supply. *Progressive Fish-Culturist* 35:66–73.

Podoll, R. T. and K. C. Irwin. 1988. Sorption of cationic oligomers on sediments. *Environmental Toxicology and Chemistry* 7:405–15.

Podoll, R. T., K. K. Irwin, and S. Bredlinger. 1987. Sorption of water-soluble oligomers on sediments. *Environmental Science and Technology* 21:562–68.

Schwoyer, W. L. K. 1981. *Polyelectrolytes for Water and Waste Water Treatment*. Boca Raton, FL: CRC Press.

Stutzmann, T. H. and B. Siffert. 1977. Contribution to the adsorption mechanism of acetamide and polyacrylamide onto clays. *Clays and Clay Minerals* 25:392–406.

Chapter 8

Dispersion Polymers

■ Patrick D. Guiney,[a] James E. McLaughlin,[b] John D. Hamilton,[a] and Kevin H. Reinert[b]
S. C. Johnson Wax[a] *and Rohm and Haas Company*[b]

Introduction

Dispersion polymers are key ingredients in the large-scale commercial formulation of paints, adhesives, floor finishes, and related products. These polymers are based primarily on acrylic chemistry. The total production volume in the United States for high molecular weight (MW) acrylic polymers for water-based products has been estimated to be 582 million kilograms per year, and acrylic polymers for coatings contribute approximately 293 million kilograms of this total (Skeist, 1994). A typical finished paint or floor finish may contain approximately 10 to 20% acrylic polymer solids. Given such high commercial-use volumes, it is relevant to discuss ecotoxicity and ecological fate of dispersion polymers as components of waste from manufacturing, formulation, product use, and disposal.

This chapter focuses on the wide variety of water-based dispersion polymers for acrylic products. Table 8.1 shows a generic product recipe with major components for a polymer dispersion that is produced by aqueous emulsion polymerization. The resulting dispersion product typically contains about 40 to 50% polymer solids. Dilute surfactant (approximately 1%) is adsorbed at the polymer–water interface to disperse the polymer solids in water. Dispersion polymer products also contain dissolved trace amounts (often less than 0.1%) of additives such as stabilization salts and/or antimicrobial biocides. Paint, adhesive, and floor finish polymer backbones can include carboxylic acids, acrylates, methacrylates, acrylamide, acrylonitrile, styrene, butadiene, and/or vinyl acetate monomers. Common monomers in acrylic coating polymers are methyl methacrylate, styrene, α-methyl

Table 8.1. Product recipe for dispersion polymer production

Component	Parts by weight
Water	180
Monomer 1 (e.g., butyl acrylate)	75
Monomer 2 (e.g., methyl methacrylate)	25
Surfactant (e.g., alkyl alcohol ethoxylate)	5
Solubilizer (e.g., pyrophosphate)	1.5
Ferrous ion/initiator (e.g., ferrous sulfate/hydroperoxide)	0.2
Alkali (e.g., sodium hydroxide)	0.1

styrene, and acrylonitrile, which provide coating film hardness, as well as butyl acrylate, ethyl acrylate, and 2-ethylhexyl acrylate, which provide coating film softness. Carboxylic acid functional groups are often added by using methacrylic acid, acrylic acid, itaconic acid, and/or maleic acid. As a class, dispersion polymers exhibit neutral (i.e., nonionic) to slightly anionic charges. However, some coatings for floor finishes and paints are cationic with the addition of nitrogen-containing functional groups (e.g., quaternary ammonium salts of alkylamino-substituted monomers) in the polymer backbone.

The MW of water-based dispersion polymers can be up to 3 million (Allen et al., 1989), but the MW values of water-based dispersion polymers for paints and adhesives are typically in the hundreds of thousands. Other coating polymers can have a MW range of approximately 1,000 to 80,000. Acrylic dispersion polymers of MW 40,000 to 60,000 serve as the main film-forming ingredient of floor finishes. Chapter 1 provides additional details on the chemistry of dispersion polymers and related technologies.

ECOLOGICAL EXPOSURE PATHWAYS

Dispersion polymers are components of finished products that are applied in household and industrial locations. The primary sources of dispersion polymers for ecological exposure are (1) industrial wastewater and sludges from polymer manufacturing, formulation, and product use; and (2) household solid waste and municipal wastewater.

Industrial wastewater from dispersion polymer production and/or product formulation locations may contain approximately 0.2 to 1.5% polymer solids. Wastewater treatment processes can recover virtually all polymer solids from wastewater prior to discharge (see below, under "Ecological Fate"). Sludges of waste paints and other coatings are also recovered by industrial users for waste handling. Polymer solids and sludges are typically disposed as stabilized solid waste. Stabilization methods include removal of water and solidification (Tucker and Carson, 1985). Minor amounts of dispersion polymers also may reach terrestrial environments from their use in dust suppression; small amounts of dilute (e.g., 1 part dispersion to 4 parts water) polymer dispersions are sprayed under limited circumstances to suppress dust on roads.

Municipal landfills and/or incinerators often are used for management of household waste (Dawson, 1979; Wilson and Rathje, 1989); so it is likely that some spent paints and other products containing dispersion polymers will reach landfills or incinerators. Further, about 15% by weight of floor finish polymers are gradually lost by floor abrasion and/or burnishing, and floor finish polymer films are eventually removed (i.e., stripped) from floor surfaces by cleaners and strippers during their normal cycle of use. The exact fractions of products reaching wastewater from the cleanup of equipment and containers are not known, but about 0.1% by weight of aqueous surfactant-based solutions remain in containers after pouring (U.S. EPA, 1990). Comparable fractions may reach municipal wastewater from the household cleaning of containers for paints, adhesives, or floor finishes. Household wash water is often discharged to municipal wastewater that reaches publicly owned wastewater treatment works (POTWs).

ECOLOGICAL FATE

General Aspects

The ecological fate of dispersion polymers in various environment compartments is driven primarily by their physical-chemical characteristics. High MW acrylic polymers tend not to biodegrade significantly, and will not be readily taken up by aquatic organisms. Estimated Henry's law coefficients for typical dispersion polymers are in the range of 10^{-16} atm-m^3/mole, indicating that these polymers will not volatilize from water. They do, however, exhibit a strong potential to sorb to soils, sludges, and sediments.

Solubility

Guidelines are being developed to standardize the measurement of polymer extractability into water (OECD, 1995). Emulsion polymerization results in a mixture of molecular sizes (polydispersity) for a given polymer backbone, and low molecular weight polymers within the mixture are potentially extractable into water. The extraction of polymers into water provides an estimate of polymer solubility at equilibrium.

Most dispersion polymers are prepared in the dispersed state. Even in the presence of surfactants, dispersion polymers are virtually (i.e., >99.95%) insoluble in water. When placed in dispersant-free water, dispersion polymers will destablilize to form filterable aggregates. Lack of polymer solubility in wastewater gives a basis for removing suspended polymers from wastewater by filtration and/or coagulation prior to discharge.

Sorption and Partitioning

Dispersion polymers will bind tightly to organics found within soils and sediments. The sorption partition coefficient (K_d) describes the affinity of a polymer for a sorbent (e.g., sludge biomass, sediment, or soil). Sorption of coating polymers, for example, in various ecological compartments can be described by a simple linear model in which mass or concentration of a polymer on a sorbent solids (mg/g) is related to its concentration in solution (mg/mL) by its sorption coefficient K_d (Larson and Vashon, 1983):

$$K_d = \frac{C_{\text{solids}}}{C_{\text{solution}}}$$

where:

C_{solids} = mg polymer/g on sorbent solids

C_{solution} = mg polymer/mL in solution

Assumptions for using the above relationship are that (1) organic matter is the sorbent, (2) there are no unique reactions between the sorbent and the polymer, (3) sorption is instantaneous, (4) equilibrium is achieved, (5) sorption may be reversible, and (6) the sorption isotherm is linear (Scow et al., 1995). These assumptions are valid for the many dispersion polymers that do not contain polar functional groups capable of forming chemical bonds with sorbent organic matter.

Figure 8.1 illustrates a linear Freundlich adsorption isotherm on sludge biomass for a typical dispersion polymer for floor finishes (Jop et al., 1996). The dispersion polymer (designated PE-41) has a MW of 40,000 to 50,000 with an anionic backbone of styrene, α-methyl styrene, methyl methacrylate, butyl acrylate, and methacrylic acid. PE-41 exhibited a relatively high K_d value of 745, indicative of a strong potential to sorb to organic matter both in the natural environment and in wastewater. To standardize sorption data, the concentration of sorbed polymer can be expressed on an organic carbon basis (K_{oc}); adsorption is correlated with organic carbon content. For approximately 27% sludge organic carbon content, K_{oc} was 2,730. Compounds having K_{oc} values greater than 1,000 are tightly bound to organic matter and are considered virtually immobile.

The degree of sorption can influence the route, the extent of exposure, and consequently the toxicity of a chemical. For example, the presence of sediment will significantly alter the bioavailability of surfactants to sediment-dwelling species such as midge larvae (Pittinger et al., 1989). Although significant dietary contributions of bioaccumulated chemicals have been observed with some organic compounds (Landrum and Scavia, 1983), dissolved fractions of most hydrophobic chemicals appear to be more

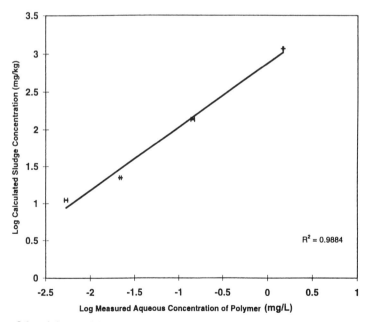

Figure 8.1. Advanced isotherm regression plot for adsorption of styrene–acrylic dispersion polymer PE-41 to POTW sludge.

bioavailable to aquatic organisms than sorbed fractions (Adams et al., 1985). As dispersion polymers tend to be tightly bound to solids and are not susceptible to biodegradation, sediments represent a potential sink (i.e., compartment) in the aquatic environment for these materials. However, it should be noted that primary routes of uptake by sediment species involve membrane absorption of dissolved compounds from interstitial water or ingestion of compounds that are adsorbed on organic particles. Primarily because of their high MW which limits uptake and their high sorption potential which reduces bioavailability, dispersion polymers are not expected to significantly impact sediment organisms (see below, under "Ecotoxicity").

Abiotic Degradation

Dispersion polymers are often designed to be resistant to abiotic oxidation and other forms of chemical degradation. The ester groups, for example, of acrylic polymers cannot be hydrolyzed under typical environmental conditions. Photodegradation of certain types of polymers (e.g., polyethylene, polypropylene, and polystyrene) containing carbonyl groups or metal-additive auto-oxidants has been observed (Barenberg et al., 1990), but comparable photodegradation of acrylic polymers has not been reported.

Biodegradation and Wastewater Simulation Tests

Physical-chemical properties of dispersible coating polymers preclude them from undergoing significant biodegradation. For a polymer to be biodegraded, appropriate microorganisms must gain access to the polymer chain and initiate an enzyme-catalyzed hydrolysis, causing chain scission with a resulting change in the molecular weight distribution. Characteristics that promote biodegradability include high water solubility, amorphous structure, and flexible chains. Biodegradation potential is also highly dependent on the molecular weight of the polymer. Acrylic oligomers do not biodegrade at MW greater than approximately 500 (Kawai, 1992).

The high degree of sorption that is characteristic of high MW acrylic polymers is another factor that limits biodegradation potential. Dissolved phases of chemicals in sediment are generally considered to be more bioavailable for biodegradation than sorbed phases (Rittman et al., 1992). Further, the longer the period of time that floor finish polymers have been in contact with sewage sludge, the more difficult it is to extract them (Guiney, 1994). Although the mechanism has not been explained, this phenomenon is consistent with the time-dependent reduction in bioavailability and degradability that has been observed for other sorbed organic chemicals (Steinberg et al., 1987).

In a preliminary program to evaluate dispersion polymer (as product containing approximately 50% polymer solids) biodegradability, standard screening methods (APHA, 1992) were used to compare biochemical oxygen demand (BOD) to theoretical chemical oxygen demand (COD) of dispersion polymer products (Rohm and Haas, 1995a). The BOD/COD ratios were substantially less than 0.50, indicative of a low potential for biodegradation. The source of limited BOD was probably product additives such as surfactants. Surfactants have shown partial or complete biodegradability (Struijs and Stoltencamp, 1994). No trend for polymer structure-related effects on BOD/COD was evident.

To obtain a realistic assessment of biodegradability, a Modified Zahn-Wallens test from Organisation for Economic Co-operation and Development guidelines (OECD 302B) was used to assess the removal potential of six water-based acrylic polymer dispersions in an inoculum of sewage sludge microorganisms (Rohm and Haas, 1995b). OECD 302B is designed to measure inherent biodegradation and sorption potential. Polymer solids were collected by filtration, and the filtered fraction (i.e., filtrate) was analyzed for total dissolved organic carbon (DOC). Total DOC was from both "filterable" polymer (i.e., the surfactant-dispersed fraction) and "dissolved" polymer (i.e., the water-soluble fraction). As shown in Table 8.2,

Table 8.2. Acrylic dispersion polymers: Estimation of wastewater fate and inhibition potential

Polymer backbone[a]	Elimination[b]	Sludge inhibition[c,e]	Microtox[R][d,e]
EA/MMA/MOLAM	Complete	$EC_{50} > 100$ mg/L	327 mg/L
EA/MMA	74%	$EC_{50} > 100$ mg/L	19,734 mg/L
BA/MMA	99%	$EC_{50} > 100$ mg/L	16,207 mg/L
BMA/EHA/MAA/MMA/STY	89%	$EC_{50} > 100$ mg/L	113,575 mg/L
BMA/MMA/STY	37%	$EC_{50} > 100$ mg/L	824 mg/L
BA/MAA	79%	$EC_{50} > 100$ mg/L	122,012 mg/L

[a]BA, butyl acrylate; BMA, butyl methacrylate; EA, ethyl acrylate; EHA, ethyl hydroxyacrylate; MAA, methacrylic acid; MMA, methyl methacrylate; MOLAM, methylolacrylamide; STY, styrene.

[b]OECD 302B test results; percent elimination of total dissolved organic carbon (DOC) from wastewater test medium (Rohm and Haas, 1995b).

[c]Aerobic activated sludge inoculum inhibition; 100 mg/L was highest dose tested.

[d]Microtox[R] assay for bacterial inhibition; 15-minute EC_{50}.

[e]All results reported here according to mg/L of *product*. Products tested here consisted of approximately 50% by weight polymer active in aqueous dispersant.

significant removal of polymer solids was observed. Sorption, not biodegradation, was the primary mechanism of removal. The phase stability of a dispersion polymer is dependent on a suitable dispersant. Dilution by the addition of water-based dispersions to the test inoculum probably caused dispersion destabilization, aggregation, and/or adsorption of the polymer particles to the inoculum sludge and filter paper. Reduction of the surfactant concentration can cause a rapid increase in dispersion polymer particle size (Chellappa, 1995). Removal occurred by sorption of polymer solids to the filter and inoculum sludge. Consistent with that study, a German study based on OECD 302B methods has reported that the mechanism of removal of acrylic polymers from wastewater is polymer sorption to inoculum sludge (Bauer, 1992).

The range in percent elimination from 37% to complete removal with the OECD 302B tests with the six water-based polymer dispersions (Table 8.2) may reflect variability in polymer aggregation and/or adsorption potential. Adsorption potential of polymers on clays, for example, has been found to decrease in the order of cationic > nonionic > anionic charges (Ben-Hur et al., 1992). However, although lower percent elimination (i.e., 37%) was predicted with the styrene–acrylic polymer with the OECD 302B test, a comparable styrene–acrylic polymer (PE-41) was virtually completely eliminated in more advanced testing. The more advanced testing included the semi-continuous Activated Sludge (SCAS) test and field sampling at a wastewater treatment plant (see below).

Biodegradation of nonacrylic dispersion coatings may be observed, but only under active culture conditions and after a long incubation period. Approximately 25% polymer weight loss after 12 weeks was found when a dispersion polymer containing vinyl acetate (VAC) and vinyl alcohol (VOH) was incubated in natural soil samples (Kastien and Sutter, 1992). Under the same conditions, no weight loss of a butyl acrylate/methyl methacrylate polymer was observed.

Even if it is assumed that a small fraction of dispersion polymers is truly water-soluble and partially biodegradable, OECD ready biodegradability "pass" criteria include 60% percent biodegradation after only 4 weeks (28 days) of incubation using highly dilute inoculum. Biodegradation results described above (e.g., BOD/COD ratios <0.5) indicate that such screening-level criteria will not be reached under test conditions for ready biodegradability. Resistance of dispersion polymers to biodegradation is associated with water insolubility, high molecular weight, large polymer size, and steric hindrance (Cooke, 1990; Huang et al., 1990). Sorption to organics is a much more important mitigation mechanism in the natural environment than in water treatment processes. Water treatment by filtration and/or aggregation of polymer solids is more realistic than biodegradation for wastewater that contains dispersion polymers.

As mentioned above, manufacturing processes and consumer use patterns can discharge products containing dispersion polymers (e.g., paints, floor finishes) to wastewater. The Semi-continuous Activated Sludge (SCAS) test can be used to estimate the extent to which a polymer can be removed in typical aerobic wastewater treatment processes. In an SCAS test with PE-41 (styrene/acrylic polymer dispersion described previously) at a maximum expected wastewater concentration (i.e., 10 mg/L as polymer), the percent removal by sorption to sludge solids ranged from 96% to 99% after 60 days (Jop et al., 1996). The metabolic activity of the SCAS aerobic digester was not adversely affected by the polymer. Physical-chemical characteristics that probably influenced the polymer's high removal efficiency were its high K_d and its tendency to precipitate with calcium.

Wastewater Treatment

The SCAS results have been recently confirmed with data obtained from field studies with PE-41 in a municipal wastewater treatment plant using aerobic sewage treatment (Guiney et al., 1996a). Based on field measurements, a combination of primary and secondary wastewater treatment removed greater than 99% of the polymer by sorption to sludge solids. PE-41 had no adverse effects on the settling behavior of sludge during primary and final sedimentation processes. The SCAS and field results indicated that removal of dispersion polymers from water will take place largely by polymer sorption to the sludge biomass in aerobic sewage treatment. Only a very small amount will undergo biodegradation.

Coagulation of polymer solids by adjustment to alkaline pH and the addition of water treatment chemicals such as ferric chloride ($FeCl_3$), followed by solids settling or filtration, provide highly effective physical-chemical methods of treating dispersion wastewater (Rohm and Haas, 1983). Approximately 98 to 99% of suspended solids and 70 to 86% of chemical oxygen demand (COD) were removed from wastewater containing polyvinyl acetate or acrylic polymers when treated with combined lime ($Ca(OH)_2$) and alum ($Al_2(SO_4)_3$) (Williams, 1974). COD sources included biodegradable organic constituents such as surfactants as well as the paint polymers. The removal effectiveness of activated sludge was evaluated in the same study. Activated sludge was less effective overall than combined chemical (i.e., lime and alum) coagulation, but gave better removal of COD when compared to lime or alum applied alone. The results indicated that a combination of coagulation as primary treatment and activated sludge as secondary treatment can be highly effective in removing both suspended solids and COD of dispersions from wastewater.

Landfill Disposal

Virtually all municipal landfills are primarily anaerobic environments, with aerobic processes found primarily at the landfill surface (Stegmann et al., 1993). However, sufficient anaerobic biodegradation occurs in relatively few landfills (e.g., in about 100 of 16,000 U.S. landfills) to produce commercial quantities of methane (Valenti, 1992). Solid waste sampling studies have shown that materials that biodegrade under natural surface conditions (e.g., food waste) can persist for years while buried in some municipal landfills (Rathje, 1989). As indicated above, dispersion polymers are resistant to biodegradation and will be virtually immobile when bound to organics. Leaching of dispersion polymer fractions is not anticipated. Dispersion polymer waste is often stabilized prior to landfill disposal; stabilization includes removal of water from the dispersions to prevent leachate contamination.

There are many studies on the effects of microbial organisms (molds and fungi) on acrylic resins for surface coatings with resultant deterioration such as product discoloration and fragmentation (Jakubowski et al., 1983; Pankhurst, 1973). However, as noted above, acrylics are resistant to biodegradation. A recent study has assessed the degradation of a butyl acrylate/methyl methacrylate dispersion polymer in soil (Kastein and Sutter, 1992). After laboratory incubation of polymer in soil at 28°C for 12 weeks, there was no evidence that significant biodegradation had occurred.

ECOTOXICITY

Activated Sludge Microbial Toxicity

Concentrations of approximately 200 to 300 mg/L of acrylic dispersion polymers for floor finishes are required to see a 50% reduction in glucose uptake (i.e., a measure of aerobic metabolism) rate by unacclimated activated sludge (Guiney et al., 1996b). The glucose uptake test was based on a variation of the standard biochemical oxygen demand (BOD) test, also known as the BOD microbial toxicity test of BOD_m (Marks, 1973). The No Observed Adverse Effect Concentration (NOAEC) for the polymers was 60 to 100 mg/L. At maximum projected wastewater discharge rates, floor finish polymers will not interfere with aerobic treatment processes after entering municipal wastewater treatment plants (Jop et al., 1996).

The aerobic sludge inhibition results shown in Table 8.2 with six water-based dispersion polymers (Rohm and Haas, 1995b) are consistent with those obtained for the floor finish polymers. The polymers were selected to represent a range of polymer backbones. Demonstrating EC_{50} values substantially greater than $100\,mg/L$, Microtox® studies suggest that the six water-based dispersions will be of low concern for aquatic toxicity. The studies suggested that high levels of acrylic polymers in wastewater will not inhibit aerobic sludge processes in wastewater treatment systems.

Aquatic Organisms

Monomer reactivity during polymerization varies according to monomer structure. The greater the tendency of a monomer is to undergo copolymerization with other monomers, the higher its reactivity. Both polymer charge and component monomer reactivity influence aquatic toxicity. Aquatic toxicity studies with over 20 dispersion polymers indicated that polymer backbones containing either monomers with nonreactive functional groups or monomers with reactive functional groups other than cationic groups (i.e., methylolacrylamide) are of low concern for aquatic toxicity (Rohm and Haas, 1995b). As shown in Table 8.3, key results were:

- Polymers shown to be of low concern for aquatic toxicity contained carboxylic acids, acrylates, methacrylates, acrylamide, acrylonitrile, styrene, butadiene, and vinyl acetate monomers.
- Polymers with reactive cationic nitrogen-containing functional groups (e.g., quaternary ammonium salts of alkylamino-substituted monomers) may be of higher concern for aquatic toxicity than those listed above.
- Polymers associated with long-chain ethoxylated amines as cationic counterions may be of higher concern for aquatic toxicity than dispersion polymers associated with ammonium cations or monoalcohol amines as counterions.

The above findings are consistent with the hypothesis that reactive monomers (and reactive counterions) may increase aquatic toxicity of dispersion polymers, at least under laboratory test conditions. However, component monomer reactivity will mitigate the potential for exposure under natural conditions. Dispersion polymers that contain reactive components are often designed to bind strongly to natural substrates as coatings. In natural environments, reactive polymers will tend to strongly sorb to available organics, thereby reducing exposure to aquatic species.

As indicated above, dispersion polymers are dispersed in an aqueous phase with the aid of a surfactant. According to tests with nine acrylic dispersion polymers, there was no influence of dispersion surfactant type on

Table 8.3. Acrylic dispersion polymers: Aquatic toxicity

Backbone	Monomer composition	Test(s)	Results[a]
Acrylic polymers with nonreactive monomers[b]	Nonreactive monomers (e.g., EA, MMA, STY, MAA, EHA)	$Daphnia\ magna$ 48-hr EC_{50} Trout 96-hr LC_{50} Algae 72-hr EC_{50}	100 to >1,000 mg/L 100 to >1,000 mg/L >100 mg/L
Acrylic polymers with cationic counterions[c]	Long-chain ethoxylated amine as counterion, plus nonreactive monomers	Trout 96-hr LC_{50} $Daphnia\ magna$ 48-hr EC_{50}	31 mg/L 81 mg/L
	Monoalcohol amine as counterion, plus nonreactive monomers	Fathead minnow 96-hr LC_{50} $Daphnia\ magna$ 48-hr LC_{50} Algae 96-hr EC_{50}	>1,000 mg/L >1,000 mg/L >1,000 mg/L
Acrylic polymers with reactive cationic nitrogen-containing monomers[d]	Alkylamino-substituted methacrylates plus nonreactive monomers	$Daphnia\ magna$ 48-hr LC_{50} Fathead minnow 96-hr LC_{50}	6.5 to 220 mg/L 4.7 to 5.3 mg/L
See above[e]	See above	Bluegill sunfish 96-hr LC_{50} Algae 96-hr EC_{50}	36 mg/L 3.9 mg/L

[a]All results reported here according to mg/L of *product*. Products tested here consisted of approximately 50% by weight polymer active in aqueous dispersant. EA, ethyl acrylate; MMA, methyl methacrylate; STY, styrene; MAA, methacrylic acid; EHA, ethyl hydroxyacrylate.

[b]17 products tested; range of results reported where available.

[c]2 products tested.

[d]3 products tested, each contained quaternary ammonium salts of alkylamino-substituted methacrylates making up ≤10% by weight of the polymer backbones; range of results reported.

[e]1 product tested; product contained quaternary ammonium salt of an alkylamino-substituted methacrylate making up ≤10% by weight of the polymer backbone.

acute aquatic toxicity (Rohm and Haas, 1989). All dispersion polymers presented LC_{50} (or EC_{50}) values of greater than 100 mg/L in algae, *Daphnia magna*, and fish. Concentrations of surfactant(s) used in polymer dispersions are often approximately 1%, indicating that overall product ecotoxicity will not be influenced by surfactant type. For example, the effluent toxicity of wastewater will not be significantly influenced by the type of dispersion surfactant.

Based on the aquatic toxicity data shown in Table 8.4, it can be concluded that styrene–acrylic dispersion polymer PE-41 was generally of low toxicity to aquatic organisms except for *Ceriodaphnia dubia* (Guiney and Jop, 1994). As indicated in Chapter 7, polymers with high adsorption potential can sorb to aquatic species during laboratory testing, possibly causing lethality by physical effects. Under test conditions with low organics present, polymer sorption to *C. dubia* may have reduced integument membrane transport or exchange of nutrients and gases (e.g., oxygen and carbon dioxide) and/or inhibited appendages that would normally facilitate gas and nutrient uptake. Higher susceptibility of *C. dubia* has been reported with other polymers (see Chapter 7), and the susceptibility of *C. dubia* has been attributed to the relatively high available body surface to volume ratio of these organisms when compared to other aquatic test organisms.

Table 8.4. Styrene–acrylic dispersion polymer (PE-41) — acute and chronic aquatic toxicity

Test	Species	Response	Polymer emulsion[a,b] (mg/L)
Algal toxicity	*Selenastrum capricornutum*	EC_{50} NOEC	100[c] 29[d]
Invertebrate toxicity	*Ceriodaphnia dubia*	48-hr NOAEC	910
	Ceriodaphnia dubia	Chronic NOEC	1[e]
	Chironomus riparius	Acute NOAEC	690
	Chironomus riparius	Chronic NOEC	718
Fish toxicity	*Pimephales promelas*	96-hr NOAEC	340
	Pimephales promelas	Chronic NOEC	680

[a]Anionic styrene–acrylic dispersion polymer with a molecular weight range of 50,000 to 60,000. Product tested was approximately 50% by weight polymer active in aqueous dispersant.

[b]All results reported here according to mg/L of *polymer*.

[c]EC_{50} of 100 mg/L at a hardness of 32 mg $CaCO_3$/L was mitigated to an EC_{50} of 1,200 mg/L at a hardness of 140 mg $CaCO_3$/L.

[d]NOEC of 29 mg/L at a hardness of 32 mg $CaCO_3$/L was mitigated to an NOEC of 500 mg/L at a hardness of 140 mg $CaCO_3$/L.

[e]Caused by physical effects; see text for discussion of ecological significance.

Elevated acute to chronic toxicity ratios often suggest a time-dependent (i.e., slower-acting) mechanism of action. Under typical *C. dubia* toxicity test conditions in the laboratory, polymer dispersions can exhibit relatively high acute to chronic toxicity ratios (i.e., 26 to 910) that may reasonably be attributed to physical effects of these polymers at test concentrations rather than direct systemic toxicity (Guiney et al., 1996b). In the natural environment, competition of the polymer for suspended organic material will probably mitigate toxicity with aquatic species such as *C. dubia*.

From the baseline set of ecological effects data presented in Table 8.4, green algae (*Selenastrum capricornutum*) were the next most sensitive aquatic species to the polymer dispersion. The primary mechanism of action may be polymer-related chelation of essential nutrients (e.g., calcium and magnesium) required by algae for growth (Hamilton et al., 1994). Increasing the hardness of the test medium from 32 mg $CaCO_3$/L to 140 mg $CaCO_3$/L decreased the algal EC_{50} and NOEC 12 times and 17 times, respectively.

Plants, Sediment, and Soil Organisms

At levels presently tested, long-term retention of dispersion polymers by adsorption to soil or landfill organics is not expected to adversely affect plants, sediment, and soil organisms. The results of acute toxicity test with midge larvae (*C. riparius*) exposed to PE-41 confirm predictions that the high MW of acrylic dispersion polymers will mitigate concern for toxicity in sediment organisms. No reduction of growth or survival occurred up to 690 mg/kg, and a chronic NOEC of >718 mg/kg was found (Jop and Guiney, 1996). Further, when PE-41 was studied with exposure to earthworms, the polymer exhibited an acute LC_{50} of 510 mg/kg, a chronic No Observed Effect Concentration (NOEC) of 250 mg/kg, and a chronic Lowest Observed Effect Concentration (LOEC) of 830 mg/kg. The measured bioconcentration factor for the dispersion polymer was less than 0.1, indicating negligible bioaccumulation potential (Jop and Guiney, 1996). Based on this chronic toxicity and bioconcentration information, dispersion polymers do not appear to present a significant chronic hazard to sediment and soil organisms.

After reaching wastewater treatment and sorbing to sludge biomass, dispersion polymers may be transferred with the biomass on land for disposal or soil amendment. Seedling emergence and growth of terrestrial plants have been evaluated with sludge biomass obtained from an SCAS test unit that was in contact for 60 days with increasing concentrations of 3 to 6 mg/g PE-41 as measured in sludge. The plant species (i.e., corn, rye grass, cucumber, lettuce, and radish) were not affected at concentrations as high as 72 mg/kg of sludge in soil (Jop and Guiney, 1996). It is interesting to note that the control (i.e., nonpolymer) SCAS unit sludge was toxic to the plants, possibly because of the elevated concentrations of toxic metals in the sludge. Addition of PE-41 to the SCAS unit reduced sludge toxicity by 50%, probably by binding to metals in the sludge. Although polymer effects on contaminant bioavailability have not been investigated directly, the potential of acrylic polymers to reduce mobility of pesticides has been observed. For example, experiments conducted with soil-applied herbicides in citrus groves of Florida showed significant reduction in the leachate mobility of bromacil, diuron, norflurazon, and simazine in soil after addition of an acrylic polymer (Reddy and Singh, 1993).

EXPOSURE MITIGATION

Adsorption to organic solids and the tendency of dispersions to form insoluble polymer solids strongly reduce the potential for systemic uptake by organisms. Formation of visible particulates by aggregation during testing is consistently observed in aquatic toxicity tests with polymer dispersions. Further, the high MW of dispersion polymers reduces the potential for absorption of polymers by organisms. The uptake of synthetic polymers by microorganisms decreases rapidly at MW more than 500 (Potts et al., 1972), and passive diffusion across gill membranes may be limited to MW of 1,000 or less (Spacie and Hamelink, 1985).

CHARACTERIZATION OF EXPOSURE

Current analytical difficulties in detecting acrylic polymers in the environment include analytical interferences with natural organic matter. An analytical method for water treatment polymers based on size exclusion chromatography (SEC) has been developed, but this method was successful only with linear nonionic and anionic polymers of very high (above one million) molecular weight (Soponkanaporn and Gehr, 1989). However, the need for advancement in low level detection (i.e., less than 1 mg/L) of dispersion polymers is not urgent. For example, soil and wastewater concentrations of an acrylic dispersion polymer (PE-41) have been estimated (Guiney and Jop, 1994; Guiney et al., 1996a), suggesting wide margins of safety between chronic NOEC values and concentrations in aquatic and terrestrial environments.

CONCLUSIONS

Based on available information, the use of dispersion polymers in products such as paints, adhesives, and floor finishes does not pose significant risk to the environment. At maximum projected use rates, dispersion polymers will not interfere with aerobic wastewater treatment processes. Dispersion polymers are not susceptible to biodegradation, but will tightly bind to soil, sludge, and sediment. On the basis of field studies in wastewater treatment facilities and SCAS tests, over 96% removal by sorption mechanisms is expected.

REFERENCES

Adams, W. J., R. A. Kimerle, and R. G. Mosher. 1985. Aquatic safety assessment of chemicals sorbed to sediments, in *Aquatic Toxicology and Hazard Assessment*, ASTM STP 854, ed. R. D. Cardwell, R. Purdy, and R. C. Bahner. Philadelphia, PA: American Society for Testing Materials.

Allen, G., J. C. Bevington, and G. C. Eastmond. 1989. *Comprehensive Polymer Science: The Synthesis, Characterization, Reactions and Applications of Polymers.* New York: Pergamon Press.

APHA. 1992. *Standard Methods for the Examination of Water and Waste Water.* Washington, DC: American Public Health Association.

Barenberg, S. A., J. L. Brash, R. Narayan, and A. E. Redpath. 1990. *Degradable Materials — Perspectives, Issues and Opportunities.* Boca Raton, FL: CRC Press.

Bauer, H. 1992. Modified Zahn-Wallens test for the determination of the eliminability of sizing polymers. *Melliand Textilberichte* 73:755–56.

Ben-Hur, M., J. Faris, M. Malik, and J. Letey. 1992. Adsorption of polymers on clays as affected by clay charge and structure, polymer properties, and water quality. *Soil Science* 153:349–56.

Chellappa, C. 1995. Tailoring particle size and stability in latex dispersions. *Modern Paint and Coatings* (Dec.):28–32.

Cooke, T. F. 1990. Biodegradability of polymers and fibers — a review of the literature, *Journal of Polymer Engineering* 9:171–81.

Dawson, R. A. 1979. Sludge management in the paint and coatings industry. *Sludge* (Sept.–Oct.):32–38.

Guiney, P. D. 1994. Environmental Extraction Analysis of PE-41. Unpublished report. Racine, WI: S. C. Johnson and Son.

Guiney, P. D. and K. M. Jop. 1994. A comprehensive ecological risk assessment of two synthetic polymers, in *Proceedings of the Society of Environmental Toxicology and Chemistry Annual Meeting*, Vol. 441, p. 80. Denver, CO: Society of Environmental Toxicology and Chemistry.

Guiney, P. D., D. M. Woltering, and K. M. Jop. 1996a. A comprehensive ecological risk assessment of two synthetic polymers. *Environmental Toxicology and Chemistry* (submitted publication).

Guiney, P. D., K. M. Jop, and K. Christensen. 1996b. Ecotoxicological assessment of two polymers of distinctively different molecular weights. *Environmental Toxicology and Chemistry* (submitted publication).

Hamilton, J. D., K. H. Reinert, and M. B. Freeman. 1994. Aquatic risk assessment of polymers. *Environmental Science and Technology* 28:186A–192A.

Huang, J.-C., A. S. Shetty, and M.-S. Wang. 1990. Biodegradable plastics: A review. *Advances in Polymer Technology* 10:23–30.

Jakubowski, J. A., J. Gyruris, and S. L. Simpson. 1983. Microbiology of modern coating systems. *Journal of Coatings Technology* 55:49–59.

Jop, K. M. and P. D. Guiney. 1996. Ecotoxicological assessment of two polymers of distinctively different molecular weights. *Environmental Toxicology and Chemistry* (submitted publication).

Jop, K. M., K. Christensen, E. Silberhorn, and P. D. Guiney. 1996. Environment fate assessment of two synthetic polycarboxylate polymers. *Ecotoxicology and Environmental Safety* (submitted publication).

Kastein, H. and H. P. Sutter. 1992. The quantitative microbiological degradation of synthetic resins and polymer dispersion for paints. *Farb. Lack.* 98:505–10.

Kawai, F. 1992. Bacterial degradation of acrylic oligomers and polymers. *Applied Microbiology and Biotechnology* 39:382–85.

Landrum, P. F. and D. Scavia. 1983. Influence of sediment on anthracene uptake, depuration, and biotransformation by the amphipod *Hyalella azteca*. *Canadian Journal of Fisheries and Aquatic Science* 40:298–305.

Larson, R. J. and R. D. Vashon. 1983. Adsorption and biodegradation of cationic surfactants in laboratory and environmental systems. *Developments in Industrial Microbiology* 24:425–34.

Marks, P. J. 1973. Microbial inhibition testing procedure, in *Biological Methods for Assessment of Water Quality*, ASTM STP 528. Philadelphia, PA: American Society for Testing and Materials.

OECD. 1995. *Draft Guideline. Solution-Extraction Behavior of Polymers in Water.* Paris, France: Organisation for Economic Co-operation and Development.

Pankhurst, E. S. 1973. Protective coatings and wrapping for buried pipes: Microbiological aspects. *Journal of Oil Colloid Chemist Association* 56:373–83.

Pittinger, C. A., D. M. Woltering, and J. A. Masters. 1989. Bioavailability of sediment-sorbed and aqueous surfactants to *Chironomus riparius* (Midge). *Environmental Toxicology and Chemistry* 8:1023–33.

Potts, J. E., R. A. Clendinning, and W. B. Ackart. 1972. *An Investigation of the Biodegradability of Packaging Plastics*, EPA-R2-046. Washington, DC: U.S. Environmental Protection Agency.

Rathje, W. L. 1989. The three faces of garbage — measurements, perceptions, behaviors. *Journal of Resource Management Technology* 7:61–71.

Reddy, K. N. and M. Singh. 1993. Effects of acrylic polymer adjuvants on leaching of bromacil, diuron, norflurazon, and simazine in soil columns. *Bulletin of Environmental Contamination and Toxicology* 50:449–57.

Rittmann, B. E., J. A. Sutfin, and B. Henry. 1992. Biodegradation and sorption properties of polydisperse acrylate polymers. *Biodegradation* 2:181–91.

Rohm and Haas Company. 1983. Dilute wastes, in *Waste Disposal of Acrylic Dispersions*, Environmental Regulatory Affairs Department. Philadelphia, PA: Rohm and Haas Company.

Rohm and Haas Company. 1989. *Review of Dispersion Polymer Ecotoxicity as a Function of Surfactant Type*, Product Integrity Department. Philadelphia, PA: Rohm and Haas Company.

Rohm and Haas Company. 1995a. *BOD and COD Data for Dispersion Products*, Environmental Regulatory Affairs Department. Bristol, PA: Rohm and Haas Company.

Rohm and Haas Company. 1995b. *Polymer Resin Products Toxicity*, Toxicology Department. Spring House, PA: Rohm and Haas Company.

Scow, K. M., S. Fan, C. Johnson, and G. M. Ma. 1995. Biodegradation of sorbed chemicals in soil. *Environmental Health Perspectives* 103:93–95.

Skeist. 1994. *Acrylic Polymers, III.* Whippany, NJ: Skeist, Inc.

Soponkanaporn, T. and R. Gehr. 1989. The degradation of polyelectrolytes in the environment: Insights provided by size exclusion chromatography measurements. *Water Science and Technology* 21:857–68.

Spacie, A. and J. L. Hamelink. 1985. Bioaccumulation, in *Fundamentals of Aquatic Toxicology*, ed. G. M. Rand and S. R. Petrocelli. New York: Hemisphere Publishing Corporation.

Stegmann, R., S. Lotter, L. King, and W. D. Hopping. 1993. Fate of an absorbent gelling material for hygiene paper products in landfill and composting. *Waste Management and Research* 11:155–70.

Steinberg, S. M., J. J. Pignatello, and B. L. Sawhney. 1987. Persistence of 1,2-dibromoethane in soils: Entrapment in intraparticle micropores. *Environmental Science and Technology* 21:1201–8.

Struijs, J. and J. Stoltencamp. 1994. Testing surfactants for ultimate biodegradability. *Chemosphere* 28:1503–23.

Tucker, S. P. and G. A. Carson. 1985. Deactivation of hazardous chemical wastes. *Environmental Science and Technology* 19:215–20.

U.S. EPA. 1990. *Preparation of Engineering Assessments*, Vol. 1, EPA Contract No. 68-D8-0112, Office of Toxic Substances. Washington, DC: U.S. Environmental Protection Agency.

Valenti, M. 1992. Tapping landfills for energy. *Mechanical Engineering* (Jan.):44–49.

Williams, R. T. 1974. Latex paint wastes — municipal charges and treatment, in *Proceedings of the 29th Industrial Waste Conference — Part One*, Engineering Extension Series No. 45. Lafayette, IN: Purdue University.

Wilson, D. C. and W. L. Rathje. 1989. Structure and dynamics of household hazardous wastes. *Journal of Resource Management and Technology* 17:200–6.

Chapter *9*

Plastics

■ R. BARTHA,[a] A. V. YABANNAVAR,[a] M. A. COLE,[b] AND J. D. HAMILTON[c]
Rutgers University,[a] University of Illinois,[b] and S. C. Johnson Wax[c]

INTRODUCTION

Virtually all commodity plastics are characterized by high molecular weight, insolubility in water, low chemical reactivity, considerable mechanical strength, and resistance to biodegradation, photolysis, and hydrolysis. These characteristics are associated with little inherent hazard in terms of ecotoxicity and safety to users of plastic products. However, irresponsible disposal of plastics as debris has triggered important ecological concerns, including the visual deterioration of public recreational areas caused by debris as well as the endangerment of marine mammals and birds by entanglement in plastic nets (Center for Marine Conservation, 1994; U.S. EPA, 1992). Plastic manufacturers have responded to such issues by developing public and industrial educational programs on environmentally responsible disposal, advocating regulatory controls to prevent the irresponsible disposal of plastics, investigating degradable plastic technologies, and supporting alternative waste management methods for plastics (APC, 1994; SPI, 1994).

Conventional means of disposal of plastic products are landfills and incineration. Waste management alternatives designed to reduce the waste inherent in landfills and incineration include source reduction through packaging efficiency, waste-to-energy incineration, recycling, and reuse. Development of plastics that degrade under environmental conditions also is intended to assist in waste management. Of particular significance from an ecological assessment perspective is the evaluation of plastics with degradation potential for their use in (1) agricultural applications where

167

they come in contact with soils and (2) disposable consumer products that are intended to degrade in municipal and household compost. The major agricultural plastics with degradability claims are modified low density polyethylene films for mulch, silage bags, and stretch wrap, as well as modified high density polyethylene containers for plant nurseries and agricultural pesticides. Plastic consumer products being investigated for degradability include utensils and packaging films based on starch–polyethylene blends, starch–polyvinyl alcohol blends, polycaprolactones, polyesters, acylated polysaccharides, and lignin–styrene copolymers.

Chapters 4 and 5 of this book outline methods to measure polymer degradation in soils and composts. The following case studies were selected to call attention to unresolved challenges as well as to highlight successful approaches in measuring the degradation of plastics in soil and compost. The approaches include some innovative adaptations that have been found appropriate to test solid, water-insoluble materials such as plastics for degradability.

CASE STUDIES — DEGRADATION IN SOIL

Screening for Biodegradability by Net CO_2 Evolution

During early stages of polymer evaluation, the exact monomer composition of plastic polymers can be proprietary, analytically difficult to confirm, and/or not cost-effective to obtain. In the absence of polymer-specific characterization data, total organic carbon (TOC) measurements and carbon dioxide (CO_2) evolution can be used to provide an initial assessment of polymer biodegradability. The following examples illustrate the range of results found in screening for biodegradability of plastics by net CO_2 evolution and loss of polymer TOC.

In a study with a plastic packaging film, cellulose served as the positive control for biodegradation (Figure 9.1). Investigators incorporated 100 mg polymer or positive control into 25 g (dry weight) of fresh loam soil (i.e., 4 mg/g). Favorable conditions for biodegradation were achieved by such steps as adjusting to neutral soil pH, controlling soil moisture to 50% of water holding capacity, and incubating with appropriate aeration at 27°C for 60 days (Pramer and Bartha, 1972). Complete conversion of 100 mg cellulose with 44% TOC would have yielded 3.66 mmol CO_2. The observed net CO_2 evolution from cellulose was 2.87 mmol (i.e., 78.4% of the theoretical maximum) in 60 days. Because initially as much as 50% of the free substrate carbon is immobilized as microbial biomass and is released only

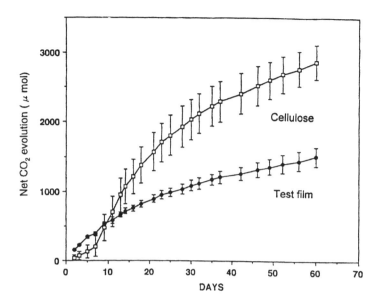

Figure 9.1.   Cumulative net CO_2 evolution in soil from a polymeric test film and cellulose (filter paper) as positive control.

at a slow rate, this result was quite typical for cellulose positive controls. The polymer TOC was 69.6%, and 100 mg could yield a theoretical maximum of 5.80 mmol CO_2. Only 1.50 mmol or 26.0% of the theoretical CO_2 was evolved. Most of this CO_2 was evolved within the first 2 weeks. Consistent with many other plastic films (Yabannavar and Bartha, 1993, 1994), the packaging film was not found to be extensively biodegradable. Substantially more biodegradation has been observed with a starch-based fill material that was developed as an alternative to styrofoam chips (Figure 9.2). The basic composition of the fill material was 85% hydroxypropylated starch, 10% formulation additives, and 5% mineral salt. After testing under the previously described conditions, 50% of polymer TOC was converted to CO_2, whereas only 2% of styrofoam TOC was converted to CO_2.

Solvent Extraction of Polymer Residues from Soil

Carbon dioxide and TOC analyses do not detect specific changes in plastics. Plastics may degrade by hydrolysis or photolysis without evidence of significant conversion of polymer TOC to CO_2. For analysis of residual sample weight or advanced analysis of polymer molecular weight and composition, samples of the polymer must be obtained. Retrieval of polymer

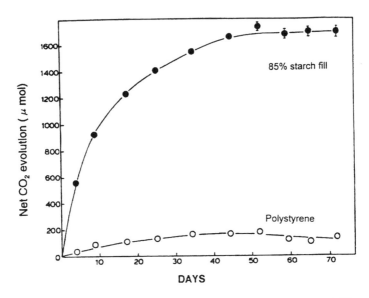

Figure 9.2. Cumulative net CO_2 evolution in soil from a starch-based fill material as compared to net CO_2 evolution in soil from conventional polystyrene foam chips.

samples by manual extraction and sample cleaning is troublesome, particularly for thin plastic films that are incubated in soil. An alternative to this method is polymer residue retrieval from soil by solvent extraction. This technique, which is commonly used in pesticide analysis, has not been widely applied to recover polymer residue samples from soil.

Solvent extraction can avoid tedious and inaccurate manual retrieval and cleaning, and allows separation of polymer residues from oxidation products and encrusted biomass. Solvent extracts may be suitable for analysis of both the plastic polymer and polymer additives by gas chromatography (GC) or high performance liquid chromatography (HPLC). For example, solvent extractions of dried soil samples containing polyvinyl chloride (PVC) have been performed in a Soxhlet apparatus using methyl ethyl ketone (MEK) as the extraction solvent (Yabannavar and Bartha, 1993). The MEK extract was split by volume in a 9 to 1 ratio. The larger portion was used after evaporation in a tared flask for determination of residual weight, and the smaller portion was used for GC analysis of PVC plasticizers.

Table 9.1 shows the zero-time recoveries of eight plasticized PVC films from soil by solvent extraction. Polymer recovery was 95% or greater, and low recovery variability was reported between sample replicates. In time-course studies, no extraction problems were encountered in the recovery of

Table 9.1. Zero-time recovery of residual weight of eight polymeric films from soil by extraction with methyl ethyl ketone (MEK)

Code[a]	Composition (% w/w)[b]	Recovery (%)[c]	SD (%)[d]
PVC-1	PVC, 69.5; DOA, 23.6; ESO, 6.9	98.4	0.5
PVC-2	PVC, 76.5; DOA, 16.0; ESO, 7.5	94.8	4.4
PVC-3	PVC, 70.0; DOA, 19.2; ESO, 10.8	103.0	1.2
PVC-4	PVC, 72.0; DOA, 22.0; ESO, 6.0	97.6	1.9
PVC-5	PVC, 70.4; DOA, 13.6; ESO, 16.0	98.0	2.4
PVC-6	PVC, 84.0; DOA, 2.0; ESO, 14.0	97.6	2.2
PVC-7	PVC, 75.4; DOA, 8.6; ESO, 16.0	97.2	2.5
PVC-8	PVC, 68.2; DOP, 24.8; ESO, 7.0	98.8	2.4

[a]Yabannavar and Bartha, 1993.

[b]PVC, polyvinyl chloride; DOA, dioctyl adipate; ESO, epoxidized soybean oil; DOP, dioctylphthalate.

[c]Percent recovery at zero-time of 250 mg polymer from 25 g soil (1% by weight).

[d]Percent standard deviation of triplicate samples.

PVC samples incubated for up to 3 months in soil that was metabolically inactivated with mercuric chloride. However, extraction methods with other polymers can be significantly different from the extraction method for PVC. MEK is not suitable for the extraction of polypropylene (PP) and polyethylene (PE). PP and PE were extracted from soil by aliquots of boiling 1,2,4-trichlorobenzene (TCB) and then precipitated by diluting the TCB with methanol (Yabannavar and Bartha, 1993). Precipitated polymer was collected on a tared filter and weighed. Although zero-time recovery of some PP and PE films by TCB extraction was comparable to the results in Table 9.1, recovery of other PP and PE films was lower (i.e., 85–90%) and more variable (Yabannavar and Bartha, 1993, 1994). In addition, PE films containing starch or insoluble pigments can yield unsatisfactory recovery results, and no solvent has provided complete extraction of light-irradiated PE and PP films. Moreover, PE films that are completely extractable by TCB at time zero can become only partially extractable after exposure to light or after incubation in soil. This alteration in extractability currently necessitates conventional manual retrieval and cleaning of the film residues (Yabannavar and Bartha, 1994).

The extraction studies illustrate a key issue in polymer analysis for degradability. Gel permeation chromatography (GPC) measurements are increasingly regarded as the most definitive proof for polymer degradability. The GPC results are used to measure weight average molecular weight (M_w), number average molecular weight (M_n), and polydispersity (Pd) of polymers, where Pd is the ratio of M_w and M_n. Definitive GPC measurements depend on complete dissolution of polymer residues in an appropriate solvent. If a portion of the cross-linked polymer residue does not dissolve, GPC results will be biased toward the soluble polymer fractions and thereby overestimate the extent of degradation. The degree of polymer residue dissolution during extraction should be reported alongside GPC results.

Issues in Measurement of Polymer Degradability

Ideally, different measurements for biodegradation should correlate when applied to the same polymer. Unfortunately, quite often this is not the case. For example:

- A substantial loss in tensile strength may be accompanied by only limited CO_2 evolution and residual weight decline, and no measurable change in polymer molecular weight as measured by GPC. In other instances, changes in polymer molecular weight are not accompanied by significant evolution of CO_2 (or vice versa). The disparity in the results is probably associated with the inherent complexity of plastics. Small changes in one component may strongly affect the mechanical properties of a plastic. Mechanical properties such as tensile strength can be strongly affected by biodegradation of "minor" plasticizer additives without significant evolution of CO_2 and changes in molecular weight.
- Additive degradation usually reduces elongation prior to breakage, as the presence of plasticizers allow the polymer strands to slip past each other. However, incubation in soil of a polyethylene film containing 7.7% starch and some pro-oxidants resulted in doubled elongation, an unexpected result (Yabannavar and Bartha, 1994). Removal of starch by biodegradation may have promoted the intramolecular slippage of the nondegradable PE strands.
- Breakage of long polymer strands by photodegradation or free radical reactions may reduce molecular size according to GPC measurements, but the reduction may not be sufficient to bring about significant biodegradation of the molecules (i.e., there may be no detectable CO_2 evolution).
- Cross-linking of plastic polymers by light irradiation may result in reduced CO_2 evolution, and cross-linking can go undetected by GPC when cross-linked polymers become insoluble in GPC solvent.

The complexities outlined above may be discouraging, but without their consideration, degradation results reported in the literature would make little sense. Examples with specific polymers follow.

Degradation Studies of Polymers with Modification

Because most intact synthetic polymers with molecular weight in excess of 100,000 are poor candidates for microbial attack, some plastic polymers are modified to enhance their susceptibility to thermal stress or light. Such modification has been evaluated for its capacity to facilitate biodegradation in a case study of two plastic films, one clear and the other black (Yabannavar and Bartha, 1994). Both films contained photosensitizers as modifiers to enhance their degradation potential. The films were exposed to sunlight for 6- and 12-week periods. After exposure, both unexposed and exposed film samples were incubated in soil.

Sunlight exposure reduced tensile strength of the clear film and led to substantial fragmentation of the black film. According to GPC, changes in M_w (decrease) and Pd (increase) were greater in the black film than in the clear film. The M_w continued to decline during 3 months of soil burial in both films. No gravimetric weight loss was observed during soil incubation, and no more than 5% conversion of TOC to CO_2 occurred. This low amount of biodegradation was not proportional to the photochemical damage (i.e., fragmentation). The film sustaining the greatest degree of fragmentation (black film, 12-week exposure) evolved the least amount of CO_2 (1.5% of TOC). Therefore, photosensitization did not appear to promote biodegradation. Further, the light irradiation period needed for extensive fragmentation, especially in the case of the clear film, was much longer than would normally occur during routine processing of municipal solid waste. However, the potential for fragmentation from long-term exposure to light may be used to reduce the risk to wildlife from contact with intact plastic as surface debris.

Degradation Studies of Polymers without Modification

Poly-3-hydroxyalkanoates (PHA) are polyesters synthesized by bacteria, which are of considerable commercial interest. For example, a 3-hydroxybutyrate/3-hydroxyvalerate copolymer is used in the manufacture of disposable shampoo bottles and other consumer products (Ogando, 1992). Tests with a 3-hydroxybutyrate homopolymer and the 3-hydroxybutyrate/3-hydroxyvalerate copolymer incubated in soil at 15, 28, and 40°C showed gravimetric weight losses in soil ranging from 0.03% to 0.64% per day (Mergaert et al., 1993). The copolymer weight loss was somewhat faster than the weight loss of the homopolymer. Molecular weight was measured in terms of M_n and M_w. No change in molecular weight was observed with GPC after soil incubation, and Pd remained essentially constant. The homopolymer and the copolymer were also incubated in sterile buffer at temperatures of up to 55°C. No gravimetric weight loss was observed in the sterile buffer, but at temperatures above 40°C, the M_w started to decline. The decline in M_w in sterile buffer suggested polymer hydrolysis at elevated temperatures. Bacteria, actinomycetes, and molds that were isolated from the soil-exposed PHA samples had the capacity to produce clearing zones on agar plates (i.e., hydrolyze PHA) in a suspended PHA overlay.

The results described above indicate that PHA polymers in soil are probably degraded by two mechanisms. One of these polymer hydrolysis, is an abiotic process that can be accelerated at elevated temperatures, as shown in the study. As an initial step PHA hydrolysis may have occurred at the surface of the soil-incorporated particles, but a loss of PHA monomers was caused by a combination of biodegradation and sample cleaning. Biodegradation was a plausible means of PHA degradation because microorganisms isolated from the soil samples were able to metabolize PHA in a PHA overlay. Nevertheless, the removal of the hydrolysis products left the GPC profile unchanged. This case study illustrates that an analytical focus on molecular weight according to GPC could have led to the erroneous conclusion that the PHA polymers are not biodegradable in soil. The study should also serve as a warning against making judgments regarding mechanisms and degree of biodegradation on the basis of one analytical method alone (e.g., GPC).

Other case studies further emphasize the need to use multiple measurements to predict overall biodegradation potential. Plasticized PVC films exposed to soil lost up to one-third of their weight, and up to 25% of polymer TOC was converted to CO_2 (Yabannavar and Bartha, 1993). The weight losses correlated to the relatively high amounts of plasticizer additives (e.g., dioctyl adipate, dioctyl phthalate, and epoxidized soybean oil) in the films. Viscosimetric and chloride release measurements gave no evidence of significant PVC degradation as polymer. The loss of plasticizers reduced the polymer elongation potential by over 90%. Studies on two PE films containing starch and proprietary pro-oxidants gave results consistent with those found with PVC films. Gravimetric weight loss and CO_2 evolution were related to the starch content of these films; these measurements were significant for the film with 7.7% starch and barely detectable for the film with 1.5% starch. Exposure in soil had minor effects on the tensile strength of the low starch film. However, such exposure more than doubled the elongation capacity of the high starch film. As indicated above, removal of starch by biodegradation may have enhanced the intramolecular slippage of nondegradable PE strands. GPC measurements showed that the M_w, M_n, and Pd of PE did not change during soil exposure.

Three PE films with high starch content (i.e., 29, 52, 67% by weight) were exposed to soil (Goheen and Wool, 1991), and gravimetric weight loss and Fourier transform, infrared (FTIR) analysis were used to monitor changes in PE during incubation. Various infrared frequencies were monitored; the decrease of carbonyl stretching at 960 to $1,190\,cm^{-1}$ measured changes in starch content. It should be noted that the very high starch (52% and 67%) films became too opaque for FTIR after 2 to 3 weeks of incubation in soil. This problem was solved by applying heat and pressure to the retrieved films, after cleaning, to restore their translucence. After 240 days of incubation of the three films in soil, the 52% and 67% PE–starch films lost 56% and 41% of their total weight, respectively. The 29% PE–starch film lost 13% of its total weight. Except for the 29% starch film, weight loss and starch degradation measured by FTIR were well correlated. Based on weight loss in the 52% and 67% PE–starch films, 11% residual starch was predicted to remain in the blends. FTIR measurements confirmed approximately 11% residual starch in the 67% and 52% films, but showed 21% residual starch in the 29% film. Further, the study gave no conclusive evidence of PE degradation as polymer although 4% weight was lost also from a 100% (nonstarch) PE film. This weight loss was ascribed to degradation and/or leaching of low molecular weight PE strands. Incomplete degradation of starch in the three PE–starch films parallels the degradation of plasticizers reported elsewhere (Yabannavar and Bartha, 1993); physical occlusion by PE of otherwise degradable material has been postulated.

CASE STUDIES — DEGRADATION IN COMPOST

Modified Polyolefin and Starch–Polyolefins

Modified polyolefin blends based on PE or PP have been studied and marketed for degradability more than any other plastic material. Processing requirements of polyolefin blends are similar to those of unmodified polyolefins, and agricultural and consumer acceptance of products that contain polyolefin blends is good. Some polyolefin blends use catalysts to accelerate polymer photodegradation and/or thermal degradation (Schwab, 1990) or incorporate a photosensitive linkage to the polymer backbone (Guillet, 1987). Catalysts used in commercial products include complexes of unsaturated fatty acids and stearate salts of a transition metal such as iron or manganese. In the presence of light or at elevated temperatures, transition metal–fatty acid complexes generate free radicals that carry out chain scission of PE and other polyolefins (Griffin, 1976; Maddever and Campbell, 1990). Chain scission mechanisms are illustrated in Figure 9.3. Other polyolefin blends are formulated by mixing a polyolefin with granular starch (Griffin, 1987). Starch degrades very rapidly in compost; complete weight loss of pure starch films in compost after 49 days has been reported (Vikman et al., 1995).

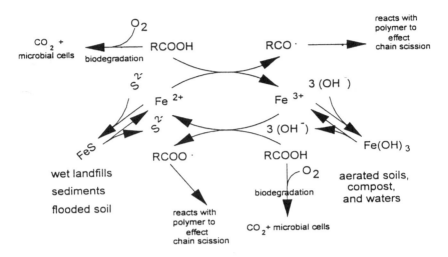

Figure 9.3. Mechanisms of polymer chain scission and biodegradation.

Results from photodegradation and biodegradation studies have illustrated the nature of polyolefin blend degradation. The percent elongation potential of PE bags containing a proprietary photosensitization additive decreased significantly when the bags were exposed for 7 weeks on the surface of compost (i.e., sunlight-exposed), but did not decrease when the bags were buried in compost (Maddever and Campbell, 1990). Intensification of carbonyl bands by FTIR occurred in surface-exposed PE but not in buried samples (Greizerstein et al., 1993). Other studies have also reported that sunlight-exposed PE–starch blends disintegrated faster than PE–starch blends that were buried in compost (Aeschleman and Cole, 1995; Cole and Leonas, 1991; Johnson et al., 1993; Leonas et al., 1994). These results indicate that PE blends are degraded faster by light than by elevated temperature (i.e., 50–60°C) in compost.

In one study, the molecular weight of several PE blends containing starch and/or transition metal catalysts and other additives was reduced after the blends were buried for one year in compost (Johnson et al., 1993). The greatest reduction in molecular weight was from 225,000 to 110,000. Other studies have not reported such a large decrease in polymer molecular weight. For example, polymer molecular weight was not reduced under aerobic or anaerobic conditions in laboratory reactors (Krupp and Jewell, 1992; Ndon et al., 1992). In addition, limited evolution of CO_2 is typically measured in plastic–starch blends, suggesting that only the starch component is degraded (Barak et al., 1991; Krupp and Jewell, 1992; Ndon et al.,

1992). The elongation, modulus, and strength of PE–starch and starch–PP films were not altered during 6 months of burial in leaf compost (Gilmore et al., 1992), indicating little change in polymer composition. Based on laboratory studies, the time to embrittlement of PE and PP films containing fatty acids and transition metals from exposure to compost at 55°C was estimated to be 2 years to greater than 70 years (Sipinen and Rutherford, 1993). Exposure at 55°C in compost thus has little effect on degradation. Effects on molecular weight are probably highly dependent on conditions of incubation and polymer composition.

Slower degradation of PE blends was found in seawater compared to degradation of the blends exposed to air (Pegram and Andrady, 1989). PE–starch films with metal catalysts were not degraded after 240 hours of exposure to ultraviolet light on the surface of salt water (Leonas and Gorden, 1993). Photodegradation of polyolefin blends in solid waste at composting facilities is not likely to be substantial; municipal waste composting usually is done indoors.

Regeneration of the reduced metal ion by redox reaction during catalysis is essential to continue polyolefin degradation, and mechanistic explanations for loss of catalysis have been sought. Figure 9.4 shows that Fe^{3+} is the thermodynamically favored oxidation state for iron under

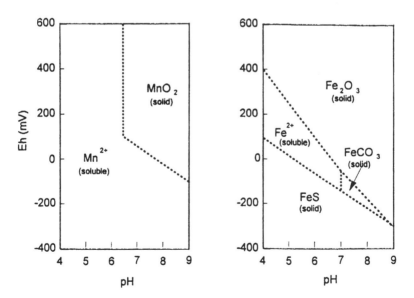

Figure 9.4. Iron and manganese speciation under commonly encountered environmental Eh and pH values (Garrels and Christ, 1965).

aerobic and alkaline conditions (positive Eh and pH > 7), whereas Fe^{2+} is favored under anaerobic and acidic conditions (negative Eh and pH < 7). Under most disposal and environmental conditions Mn^{2+} is favored, and its solubility can vary, depending on pH and Eh. It is therefore unlikely that a particular metal catalyst will perform equally well in a wide range of disposal situations. If environmental Eh/pH values are not near the iron or manganese redox transition boundaries, it is difficult for such catalysts to be recycled for further free radical generation. Some degradation may occur after disposal, but it will cease over time. Several investigators have described results that fit this pattern (Ionatti et al., 1990; Leonas et al., 1994; Maddever and Campbell, 1990).

As shown in Figure 9.5, reactions between catalyst ions and the disposal environment can control metal speciation. There is often little prospect for catalysts to be able to cycle between Fe^{2+} and Fe^{3+}. Low degradability for current iron-based systems can be expected in marine environments because Fe^{3+} will dominate under aerobic conditions. Conditions in aerobic soils are not conducive to continued catalysis of agricultural films, and virtually insoluble salt forms of metal catalysts can dominate under anaerobic conditions such as those found in landfills. Both Fe^{2+} and Fe^{3+} salts can precipitate as insoluble oxides or sulfides under environmental conditions. Catalyst insolubility will reduce the polymer degradation potential because the necessary contact of catalyst ions with fatty acids and polymer strands will be constrained by fatty acid diffusion and intramolecular motion.

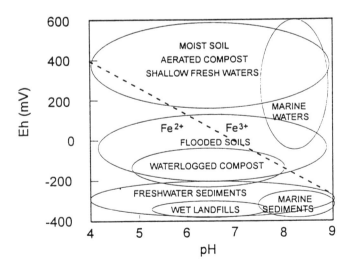

Figure 9.5.   Approximate Eh-pH relationships in disposal environments.

When precipitation is not a factor, iron or manganese from water-soluble salt catalysts has the potential to diffuse from plastics into water, thereby reducing the degradation potential. In addition, fatty acids may be lost by diffusion and biodegradation, further reducing the degradation potential. Under alkaline conditions, fatty acids exist as ionized salts of relatively high water solubility (Hansen, 1975). In environments where the pH may be greater than 8.0 (e.g., marine environments and composts), partitioning will favor migration of fatty acids to water. The fatty acid content of three PE blends with metal–fatty acid catalysts declined by 68% to 82% during composting (Chen, 1993). The loss of fatty acids will greatly decrease the capacity for free radical formation, thereby increasing the residence time of polyolefin blends in compost.

Other Polymers

Eventually, other polymers may be more likely than modified polyolefins to be completely degradable in compost. Starch–polyvinyl alcohol (starch–PVA) blends can contain 60% or more starch by weight. Some starch–PVA blends that are used as alternatives to styrofoam will hydrate rapidly and disintegrate within 48 hr of mixing with compost. About 80% of TOC from these styrofoam alternatives can be converted to CO_2 in 300 days (Bastioli et al., 1992, 1993). Up to 45% weight loss of starch–PVA blends was observed during composting (Vikman et al., 1995). Not all starch–PVA blends, however, have a high degradability potential in compost. Starch–PVA blends that are used for injection-molded products such as plastic tableware do not hydrate well in compost and are not actively colonized by microbes. Such materials have been characterized as biodegradable but not compostable.

Blends of polycaprolactone (PCL) have been produced for uses that are comparable to those of starch–PVA blends (Bastioli et al., 1995). In an study with PCL alone and PCL–PE blends incubated in compost, PCL films lost more than 95% of their initial elongation in 6 weeks (Goldberg, 1995). Blends of PCL–PE were much more resistant; for example, 80% PCL–20% PE blends lost 85% of their initial elongation only after 18 weeks. Visual inspection of starch–PCL samples reveals surface pitting and loss of components to yield a porous outer layer similar to that seen with products containing starch blended with PE or ethylene-co-acrylic acid (Cole, 1990). This appearance is typical for blends when degradation rates of components are substantially different; starch degrades faster than PCL.

The most widely studied polyesters for degradability in compost are PHA polymers such as copolymers of 3-hydroxybutyrate and 3-hydroxyvalerate. PHA polymers rapidly degrade when exposed to hydrolytic enzymes, compost, and natural sources of soil and water (Cox, 1992; Mergaert et al., 1993, 1994). Microorganisms that produce enzymes that degrade PHA are abundant in compost, and more than 90% reduction in elongation was reported after burial of PHA samples in compost for approximately 150 days (Mergaert et al., 1994).

Compost degradation of the polysaccharide cellulose acetate (CA) has been evaluated (Gu et al., 1993). Degradation was assumed to occur in the following sequence:

$$\text{cellulose acetate} \xrightarrow{\text{esterase}} \text{acetate} + \text{deacylated cellulose}$$

$$\text{deacylated cellulose} \xrightarrow{\text{cellulase}} \text{low molecular weight carbohydrates}$$

$$\text{acetate} + \text{low molecular weight carbohydrates} \rightarrow \text{microbial biomass} + CO_2$$

The results suggested that CA would be degraded rapidly in compost. However, the esterase and cellulase activities are much higher in mature composts than during earlier stages (see Chapter 5). During early phases of composting processes, undegraded materials may be screened out and discarded rather than becoming constituents of the maturing compost. In municipal composting systems, very rapid waste plastic degradation must occur; otherwise, fragments will be removed during initial screening.

Other polysaccharides are potentially more degradable than CA during early phases of composting than they are during maturation. For example, substituted cellulose acetate propionates (CAP) and blends of CAP–poly(ethylene glutarate) have been studied for degradation potential in compost (Buchanan et al., 1995). Approximately 25% to 30% weight loss was found after 30 days of composting. More rapid loss of weight occurred during the earlier thermophilic phase (62°C) than during the maturation phase (35°C). Most of the weight loss was from hydrolysis of poly(ethylene glutarate).

Lignin–styrene copolymers have also been investigated for their potential degradability under conditions that can exist in composts. Lignin–styrene copolymers contain alternating segments of pine-derived lignin and styrene. As measured by weight loss, three out of four species of white-rot fungi significantly degraded the copolymers (Milstein et al., 1994). In contrast, very little degradation of pure polystyrene occurred. White-rot fungi are commonly found in decaying plant materials and would probably be active in mature compost. When the copolymers were added to a culture medium containing white-rot fungi, the activity of lignin-degrading enzymes (i.e., lignin peroxidase and manganese peroxidase) was significantly increased. These enzymes were probably responsible for depolymerization and disintegration of the lignin–styrene copolymer.

CONCLUSIONS

The assessment of plastics for degradability has facilitated investigators' understanding of the potential for plastics to degrade when exposed to light, soils, and composts. Case studies with PVC, PHA, and PE–starch blends illustrate that multiple measurement methods should be used to verify actual degradability in the environment. For a preliminary assessment of biodegradability, CO_2 evolution and TOC measurements can be used. For an overall assessment of degradation potential, gravimetric residue weight, GC, HPLC, GPC, and/or FTIR can be used to measure polymer changes under conditions suitable for hydrolysis, photolysis, and/or biodegradation. It should be noted, however, that degradation data can be poorly correlated (e.g., CO_2 evolution versus molecular weight results by GPC). Selection of methods to measure the degradability of plastics should be based on intended degradation mechanisms. For example, rapid plastic biodegradation by microorganisms is intended for compost disposal, whereas rapid plastic fragmentation by abiotic changes in polymer composition and/or molecular weight is intended to protect wildlife from plastic debris.

As an alternative to manual retrieval and cleaning, solvent extraction is a more accurate and reliable method of sample preparation for some plastics. However, some molecular weight fractions of polyolefins can become insoluble in typical chromatography solvents because of polymer cross-linking during exposure to sunlight and/or biodegradation. Molecular weight data from GPC should not be reported as definitive unless polymer extraction was complete.

Tests in soils and compost suggests that starch alone is the biodegradable fraction of polyolefin–starch blends; PE and PP are not biodegradable under environmental conditions. Light-catalyzed polymer modifications increase susceptibility to fragmentation of plastics by exposure to light, but significant degradation will not continue when modified plastics are buried in soil or compost. Environmental conditions in compost and soils can prevent activation cycling of transition metal catalysts, reducing the capacity of modified plastics to degrade. The development of polymers such as PHA, PVA blends, polysaccharides, and lignin–styrene is continuing in studies designed to enhance the environmental degradability of plastics.

182

ECOLOGICAL ASSESSMENT OF POLYMERS

REFERENCES

Aeschleman, P. E. and M. A. Cole, 1995. Placement and seasonal effects on performance of degradable plastic bags in yard waste compost, in *Proceedings of the 37th Rocky Mountain Conference on Analytical Chemisty*, Denver, CO, July 23–27.

APC. 1994. *Use and Disposal of Plastics in Agriculture.* Washington, DC: American Plastics Council.

Barak, P., Y. Coquet, T. R. Halbach, and J. A. E. Molina. 1991. Biodegradability of polyhydroxybutyrate(co-hydroxyvalerate) and starch-incorporated polyethylene plastic films in soils. *Journal of Environmental Quality* 20:172–79.

Bastioli, C., V. Bellotti, L. Del Giudice, and G. Gilli. 1992. Microstructure and biodegradability of Mater-Bl products, in *Biodegradable Polymers and Plastics*, ed. M. Vert, J. Feijen, A. Albertsson, G. Scott, and E. Chiellini. Cambridge, UK: Royal Society of Chemistry.

Bastioli, C., V. Belotti, L. Del Giudice, and G. Gilli. 1993. Mater-Bi: Properties and biodegradability. *Journal of Environmental Polymer Degradation* 1:181–91.

Bastioli, C., A. Cerutti, I. Guanella, G. C. Romano, and M. Tosin. 1995. Physical state and biodegradation behavior of starch–polycaprolactone systems. *Journal of Environmental Polymer Degradation* 3:81–95.

Buchanan, C. M., C. N. Boggs, D. D. Dorschel, R. M. Gardner, R. J. Komarek, T. L. Watterson, and A. W. White. 1995. Composting of miscible cellulose acetate proprionate–aliphatic polyester blends. *Journal of Environmental Polymer Degradation* 3:1–11.

Center for Marine Conservation. 1994. *A Citizen's Guide to Plastics in the Ocean.* Washington, DC: Center for Marine Conservation.

Chen, C.-C. W. 1993. Mechanistic studies of biomass-filled polyethylene films degradation in refuse compost. Ph.D. dissertation, University of Missouri–Columbia, Columbia, MO.

Cole, M. A. 1990. Constraints on decay of polysaccharide–plastic blends, in *Agricultural and Synthetic Polymers: Utilization and Biodegradability*, ed. J. E. Glass and G. Swift. Washington, DC: American Chemical Society.

Cole, M. A. and K. K. Leonas. 1991. Degradability of yard waste collection bags. *Biocycle* (Mar.):56–63.

Cox, M. K. 1992. The effect of material parameters on the properties and biodegradation of "Biopol," in *Biodegradable Polymers and Plastics*, ed. M. Vert, J. Feijen, A. Albertsson, G. Scott, and E. Chiellini. Cambridge, UK: Royal Society of Chemistry.

Garrels, R. M. and Christ, C. L. 1965. *Solutions, Minerals and Equilibria.* San Francisco, CA: Freeman, Copper and Company.

Gilmore, D. F., S. Antoun, R. W. Lenz, S. Goodwin, R. Austin, and R. C. Fuller. 1992. The fate of "biodegradable" plastics in municipal leaf compost. *Journal of Industrial Microbiology* 10:199–206.

Goheen, S. M. and R. P. Wool. 1991. Degradation of polyethylene–starch blends in soil. *Journal of Applied Polymer Science* 42:2691–701.

Goldberg, D. 1995. A review of the biodegradability and utility of poly(caprolactone). *Journal of Environmental Polymer Degradation* 3:61–67.

Greizerstein, H. B., J. A. Syracuse, and P. J. Kostynaik. 1993. Degradation of starch modified polyethylene bags in a compost field study. *Polymer Degradation and Stability* 39:251–59.

Griffin, G. J. 1976. Degradation of polyethylene in compost burial. *Journal of Polymer Science* 57:281–86.

Griffin, G. J. L. 1987. Degradable plastic films, in *Degradable Plastics.* Washington, DC: Society of the Plastics Industry.

Gu, J.-D., D. Eberiel, S. P. McCarthy, and R. A. Gross. 1993. Degradation and mineralization of cellulose acetate in simulated thermophilic compost environments. *Journal of Environmental Polymer Degradation* 1:281–91.

Guillet, J. 1987. Vinyl ketone photodegradable plastics, in *Degradable Plastics.* Washington, DC: Society of the Plastics Industry.

Ionnatti, G., N. Fair, M. Tempesta, H. Neibling, F. H. Hseih, and R. Mueller. 1990. Studies on the environmental degradation of starch-based plastics, in *Degradable Materials: Perspectives, Issues, and Opportunities,* ed. S. A. Barenberg, J. L. Brash, R. Narayan, and A. E. Redpath. Boca Raton, FL: CRC Press.

Johnson, K. E., A. L. Pometto III, and Z. O. Nikolov. 1993. Degradation of degradable starch–polyethylene plastics in a compost environment. *Applied and Environmental Microbiology* 59:1155–61.

Krupp, L. R. and W. J. Jewell. 1992. Biodegradability of modified plastic films in controlled biological environments. *Environmental Science and Technology* 26:193–98.

Leonas, K. E. and R. W. Gorden. 1993. An accelerated laboratory study evaluating the disintegration rates of plastic films in simulated aquatic environments. *Journal of Environmental Polymer Degradation* 1:45–51.

Leonas, K. E., M. A. Cole, and X.-Y. Xiao. 1994. Enhanced degradable yard waste collection bag behavior in a field scale composting environment. *Journal of Environmental Polymer Degradation* 2:253–61.

Maddever, W. J. and P. D. Campbell. 1990. Modified starch based environmentally degradable plastics, in *Degradable Materials: Perspectives, Issues, and Opportunities,* ed. S. A. Barenberg, J. L. Brash, R. Narayan, and A. E. Redpath. Boca Raton, FL: CRC Press.

Mergaert, J., A. Webb, C. Anderson, A. Wouters, and J. Swings. 1993. Microbial degradation of poly(3-hydroxybutyrate) and poly(3-hydroxybutyrate-co-3-hydroxyvalerate) in soils. *Applied and Environmental Microbiology* 59:3233–38.

Mergaert, J., C. Anderson, A. Wouters, and J. Swings. 1994. Microbial degradation of poly(3-hydroxybutyrate) and poly(3-hydroxybutyrate-co-3-hydroxyvalerate) in compost. *Journal of Environmental Polymer Degradation* 2:177–83.

Milstein, O., R. Gersonde, A. Hutterman, M. J. Chen, and J. J. Meister. 1994. Rotting of thermoplastics made from lignin and styrene by white-rot basidiomycetes, in *Applied Biotechnology for Site Remediation*, ed. R. E. Hinchee, D. B. Anderson, F. B. Metting, Jr., and G. D. Sayles. Boca Raton, Fl: Lewis Publishers.

Ndon, U. J., A. D. Levine, and B. S. Bradley. 1992. Evaluation of biodegradability of starch-based plastics. *Water Science Technology* 26:2089–92.

Ogando, J. 1992. Biodegradable polymers crop up all over again. *Plastic Technology* 8:60–62.

Pegram, J. E. and A. L. Andrady. 1989. Outdoor weathering of selected polymeric materials under marine exposure conditions. *Polymer Degradation and Stability* 26:333–45.

Pramer, D. and R. Bartha. 1972. Preparation and processing of soil samples for biodegradation studies. *Environmental Letters* 2:217–24.

Schwab, F. C. 1990. Effects of photodegradants on the environmental fate of linear low density polyethylene, in *Degradable Materials, Perspectives, Issues, and Opportunities*, ed. S. A. Barenberg, J. L. Brash, R. Narayan, and A. E. Redpath. Boca Raton, Fl: CRC Press.

Sipinen, A. J. and D. R. Rutherford. 1993. A study of the oxidative degradation of polyolefins. *Journal of Environmental Polymer Degradation* 1:193–202.

SPI. 1994. *Plastics and Marine Debris — Solutions through Education, 4th Ed.* Washington, DC: The Society of the Plastics Industry, Inc.

U.S. EPA. 1992. *Plastic Pellets in the Aquatic Environment — Sources and Recommendations*, Final Report, EPA 842-B-92-010, Office of Water. Washington, DC: U.S. Environmental Protection Agency.

Vikman, M., M. Itavaara, and K. Poutanen. 1995. Measurement of the biodegradation of starch-based materials by enzymatic methods and composting. *Journal of Environment Polymer Degradation* 3:23–29.

Yabannavar, A. V. and R. Bartha. 1993. Biodegradability of some food packaging materials in soil. *Soil Biology and Biochemistry* 25:1469–75.

Yabannavar, A. V. and R. Bartha. 1994. Methods for assessment of biodegradability of plastic films in soil. *Applied and Environmental Microbiology* 60:3608–14.

PART *3* | *Environmental Regulations*

Chapter *10*

Environmental Assessment of Polymers under the U.S. Toxic Substances Control Act

■ ROBERT S. BOETHLING AND J. VINCENT NABHOLZ
  U.S. Environmental Protection Agency

INTRODUCTION

The safety of specific chemical substances in the United States is evaluated primarily under three statutes. Substances used as food additives, drugs, or cosmetics are registered by the Food and Drug Administration (FDA) under the Federal Food, Drug, and Cosmetic Act (FFDCA). Chemical substances proposed for use as pesticides are registered by the United States Environmental Protection Agency (U.S. EPA) under the Federal Insecticide, Fungicide, and Rodenticide Act (FIFRA), which imposes a host of data requirements for any submitter seeking to register the substance as an active ingredient. For the vast majority of polymers, however, the Toxic Substances Control Act (TSCA) is the applicable statute. TSCA (Public Law 94-469) was enacted by Congress in 1976, in response to a perceived need to limit exposure to industrial "environmental chemicals" such as polychlorinated biphenyls (PCBs). As stated in the Act, its primary purpose was:

> ... to assure that ... innovation and commerce in ... chemical substances and mixtures do not present an unreasonable risk of injury to health or the environment. (TSCA, Section 2(b))

As defined in TSCA, "chemical substances" specifically exclude substances already regulated under FFDCA and FIFRA (unless they have non-FFDCA or non-FIFRA uses), as well as alcohol, tobacco, and certain other materials; but all other substances are included. Thus, polymers used in water treatment, coatings, household laundry products, and manufactured goods are often subject to the requirements of TSCA.

187

TSCA requirements are different for existing substances and substances not yet in production (i.e., new chemicals). One of the first tasks of the newly created Office of Toxic Substances, now the Office of Pollution Prevention and Toxics (OPPT), was to assemble and publish a list of chemical substances already in commerce. This was accomplished in 1979 as the TSCA Chemical Substance Inventory, commonly referred to as the Inventory, which listed approximately 50,000 substances then in production or being imported into the United States. Since that time the Inventory has grown to include over 70,000 substances with the addition of new chemicals.

Anyone who wishes to manufacture or import into the United States for commercial purposes a substance not listed on the Inventory and not otherwise excluded by TSCA (e.g., pesticides and drugs) must submit formal notification of the intent to do so. Such a submission is called a Premanufacture Notice (PMN), and for most new chemicals it must be submitted to U.S. EPA at least 90 days prior to manufacture or import. New chemicals not regulated by U.S. EPA are then added to the Inventory and become existing chemicals when U.S. EPA receives a required Notice of Commencement (NOC). The NOC declares the submitter's intent to commence manufacture or import. As shown in Figure 10.1, since 1979 U.S. EPA has received more than 30,000 valid PMNs. Submissions currently average well over 2,000 per year.

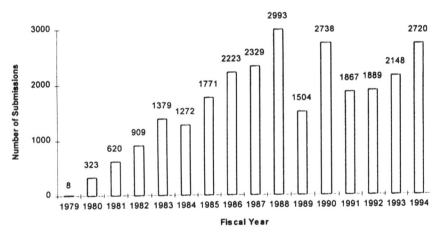

Figure 10.1. Total number of valid Premanufacture Notices received by U.S. EPA for the period 1979–1994. (Total includes PMNs, low volume exemptions, test-market exemptions, and polymer exemptions.)

THE NEW CHEMICAL REVIEW PROCESS

General Aspects

Receipt of a PMN sets in motion a review process that has evolved since July 1979 to meet the unique requirements established by TSCA. The fundamental purpose of PMN review, as stated in the law, is to determine whether:

> ... the manufacture, processing, distribution in commerce, use, or disposal [of a new chemical substance] or any combination of such activities presents or may present an unreasonable risk of injury to health or the environment .... (TSCA Section 5(b))

However, TSCA imposed serious challenges to U.S. EPA's ability to accomplish this. The most significant are, first, that PMN submitters are only required to furnish health and safety studies already in their possession (if any) and are not required to conduct any testing as a precondition for notification, and, second, that the review must be done within 90 days. The 90-day review period may be extended to 180 days under certain circumstances. If no action to regulate the substance is taken by U.S. EPA, the submitter is free to commence manufacture or import after expiration of the PMN review period. The burden of proof that a substance presents or may present an unreasonable risk rests on the Agency's shoulders. U.S. EPA must make sound decisions based on few or no submitted test data within 90 days. This situation is unlike that for pesticides in the United States and for new commercial chemicals in the European Union where specified tests to provide required data are prescribed by law.

Risk is defined as the probability of occurrence of an adverse health or ecological effect associated with exposure to the substance. OPPT's new chemical review program (NCP; U.S. EPA 1995a) is a risk assessment process. According to the U.S. National Research Council (NRC), risk assessment consists of four components: hazard (or effect) identification, dose–response assessment, exposure assessment, and risk characterization (NRC, 1994). The NCP in OPPT includes these steps, but generally does not adhere to the exact standards set by the NRC because of the aforementioned lack of data. In addition, OPPT's determination of which risks are "unreasonable" includes assessment of relative risk (i.e., comparison of relative hazards of the PMN substance and similar existing substances) and certain nonrisk factors. Foremost among nonrisk factors are those that are economic in nature, such as the costs or the benefits of the new chemical, the cost of any additional testing that may be required, the economic impact of testing or regulation on the submitter, and the pollution prevention potential (if any) associated with manufacture and use of the new chemical.

Figure 10.2 outlines the NCP in OPPT. More detailed information about the NCP can be obtained from recent publications (Moss et al., 1996; Nabholz et al., 1993a; Wagner et al., 1995). As a result of the review, a new chemical substance may be:

- Dropped from review because there is low concern about its toxicity toward humans and the environment.
- Dropped from review because there is low risk with respect to its manufacture, distribution, use, or disposal.
- Dropped from review, but with specified concerns related to its manufacture, distribution, use, or disposal, which are communicated to the submitter via letter and/or a significant new use rule (SNUR; U.S. EPA 1995b).
- Regulated, owing to either potential risk or potential significant release to the environment as a result of the initial review.
- Subjected to either a detailed review ("standard review") or an immediate request to the submitter for testing and/or additional information to determine if the substance should be regulated. Immediate testing and/or information can be requested only if the chemical belongs to a Section 5(e) chemical category (Moss et al., 1996).
- Regulated or dropped from further review as a result of standard review, additional information, and/or testing.

Section 5(f) of TSCA grants U.S. EPA the authority to take immediate action if there is a reasonable basis to conclude that a new chemical's manufacture, processing use, or disposal *will* present an unreasonable risk. However, in practice, 5(f) findings are seldom if ever made. Regulation under a Section 5(e) Consent Order, which is issued when it is found that a substance *may* present an unreasonable risk, is much more common. Through a Consent Order, the manufacturer or the importer of a new chemical consents to the order's requirements in exchange for being permitted to manufacture or import the substance. Section 5(e) findings are always predicated on insufficient information to adequately assess risk, and typically include requirements such as:

- Using protective equipment (e.g., gloves, respirators, and goggles for workers).
- Use restrictions.
- Restrictions to certain manufacturing, processing, use, and disposal sites.
- Warning labels and Material Safety Data Sheet (MSDS) statements.
- Testing to resolve uncertainties regarding toxicity or exposure.

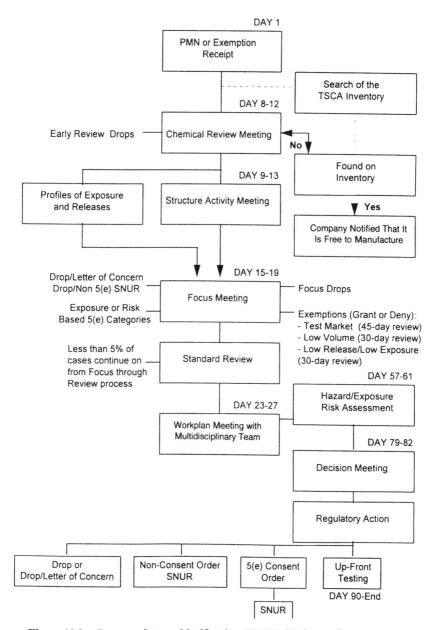

Figure 10.2. Premanufacture Notification (PMN) 90-day review process.

Testing is a frequent requirement under Section 5(e), but specific information about releases and uses is requested. If exposures can be controlled and potential risks sufficiently reduced, then the testing is generally triggered so that it is required only if and when a predetermined production volume (identified in the economic analysis) is reached. If exposures and, thus, potential risk cannot be controlled, then "up-front" testing is generally required.

Section 5(e) findings may be either risk-based or exposure-based. Prior to 1988 nearly all the regulations under 5(e) were mainly risk-based. More emphasis was placed on toxicity than on exposure in the "may present" finding. In 1988, OPPT implemented an alternative strategy, exposure-based review (XBR), that emphasized substantial or significant human exposure or substantial release to the environment. Criteria that defined the terms substantial and significant were developed for production volume, worker exposure, consumer exposure, ambient and general population exposure, and environmental release — both total releases to the environment and releases to surface waters (Table 10.1). U.S. EPA estimates these parameters on the basis of data submitted in the PMN as well as its own databases and models. If appropriate criteria are met, the outcome is usually an exposure-based 5(e) Consent Order for health effects, ecological toxicity (i.e., ecotoxicity), and/or environmental fate testing.

Full or partial exemption from review may be granted for new chemicals meeting certain requirements. U.S. EPA has limited reporting requirements for:

- Substances manufactured in small quantities solely for research and development, as long as special procedural and record-keeping requirements are met.
- Substances submitted as Test Market Exemption (TME) requests. TMEs undergo expedited (45-day) review, as the exposure assessment generally considers only the volume, number of customers, and period of time specified in the notice.
- Substances to be manufactured in quantities of 10,000 kg or less per year, which are submitted as Low Volume Exemption (LVE) requests (U.S. EPA, 1995c; see 40 CFR part 723). Low Volume PMNs undergo an abbreviated (30-day) review, and, as with TME requests, the LVE is either granted or denied. LVEs are not added to the Inventory but are maintained in a separate list.
- Substances expected to have low release to the environment and low exposures (LoREx exemption, U.S. EPA 1995c; see 40 CFR part 723), which also undergo an abbreviated (30-day) review, with the exemption either granted or denied.

Table 10.1. Criteria for substantial or significant exposure in exposure-based review of Premanufacture Notices

Type of exposure	Exposure criterion
Worker exposure: substantial or significant	
High number of workers	$\geqslant 1,000$ workers exposed
Acute worker inhalation	$\geqslant 100$ workers exposed by inhalation to $\geqslant 10$ mg/day
Chronic worker exposure	
Inhalation	$\geqslant 100$ workers exposed to $1-10$ mg/day for $\geqslant 100$ days/yr
Dermal	$\geqslant 250$ workers exposed by routine dermal contact for $\geqslant 100$ days/yr
Nonworker exposure:	
Consumer exposure via direct contact with consumer products	Presence of the chemical in any consumer product where the physical state or manner of use makes exposure likely
General population: significant exposure	
Surface drinking water	$\geqslant 70$ mg/yr
Air	$\geqslant 70$ mg/yr via ambient air
Ground drinking water	$\geqslant 70$ mg/yr
Environment: substantial exposure	
Aggregate ambient exposure through surface water, air, and groundwater	Total release to environmental media $\geqslant 10,000$ kg/yr
Substantial release to water	Release to surface water $\geqslant 1,000$ kg/yr after treatment

- Certain classes of polymers that are not chemically active or bioavailable (U.S. EPA, 1995d; see 40 CFR part 723), which are discussed in more detail in the following section.

Chemistry Review

The PMN review process can be separated conceptually into four phases: chemistry review, toxicity evaluation, exposure assessment, and risk assessment/risk management. The chemistry review phase is the first step and begins as soon as the PMN has been checked for completeness of required information, which includes all available data on chemical identity, production volume, by-products, proposed use(s), environmental release, disposal practices, and human exposure. The first step in the chemistry review is to check the adequacy of the submitted chemical name. U.S. EPA now requires the chemical name to be consistent with Chemical Abstracts Service (CAS) nomenclature policies as well as names used for similar substances already on the Inventory. Consistency between chemical name, chemical structure, and the chemical's manufacturing process is assured. A full report on the chemistry of the new substance, containing submitted information on the chemical identity, route of synthesis, impurities or by-products of the synthesis, and some physical-chemical properties, is then prepared.

For most PMNs, however, the U.S. EPA is unable to conduct a meaningful review based on submitted data alone. In large part this problem is due to TSCA's stipulation that only data "known to ... or reasonably ascertainable by" (i.e., already in the possession of) the submitter must be provided to U.S. EPA. The issue of missing data is manifested in the earliest phases of review, as even the most basic information on properties such as melting and boiling temperatures and vapor pressure often is absent. In one study of submitted PMN data, for example, only 300 chemicals with submitted data for any of several physical-chemical properties important in environmental assessment were identified from 15,000 PMNs for the period 1979–1989 (Lynch et al., 1991). Although many of the 15,000 substances were polymers, at least half of these submissions were for class 1 substances (i.e., single compounds composed of particular atoms arranged in a definite, known structure).

Polymers are class 2 substances, meaning that they have variable compositions and/or are composed of complex combinations of reactants, such that only a representative molecular structure can be drawn. Properties such as melting and boiling temperature are relevant for discrete (class 1) substances but, except for water solubility or water dispersibility, have little meaning for polymers. For polymers the chemistry review focuses on:

- The monomers of which the polymer is composed and their mole percentages.
- The molecular weight (MW) distribution, including the number average molecular weight (MW_n) in daltons, how it was determined, and the oligomer content of the polymer (i.e., percentages of oligomers with MW_n less than 1,000 and 500).
- The equivalent weight of any reactive functional groups and/or cationic charge density.

- Properties such as physical form, particle size distribution, swellability, water solubility, and water dispersibility.

Water solubility, water dispersibility, MW, charge, and cationic charge density are the most important physical properties for aquatic toxicity assessment, but basic information is often missing or is not reported in a way that is most useful for risk assessment. For example, PMN submitters are not required to report typical MW_n values, but this information is often quite useful, especially if the typical and lowest values are far apart. In addition, MW values given as greater than ($>$) or less than ($<$) an estimated MW are not helpful unless the number approximates the actual MW (e.g., "$>10,000$" is not helpful when the actual MW is 100,000). Further, monomer composition is sometimes incomplete, or chemical structural diagrams fail to show the most likely types of linkages between monomers. For example, a random reaction between monomers rather than a blocked reaction between monomers is a very important distinction that most submitters fail to make explicit. Monomer linkage is particularly important information for polymers that are claimed in the PMN to be biodegradable.

Based on its experience in reviewing over 10,000 PMNs for polymeric substances, U.S. EPA has identified a group of polymers that it believes poses low to no unreasonable risk of harm to human health or the environment, owing to, generally, their low toxicity. As of May 30, 1995 (U.S. EPA, 1995c; see 40 CFR part 723) these polymers became fully exempt from reporting. To qualify for this exemption, the polymer (U.S. EPA, 1996) must:

- Belong to one of 12 acceptable polymer classes: polyesters, polyamides and imides, polyacrylates, polyurethanes and ureas, polyolefins, aromatic polysulfones, polyethers, polysiloxanes, polyketones, aromatic polythioethers, polymeric hydrocarbons, and phenol–formaldehyde copolymers.
- Have oligomer content less than 25% by weight below 1,000 MW *and* less than 10% by weight below 500.
- Have no more than the permissible level of cationic character, which is a functional group equivalent weight for cationic groups $\geqslant 5,000$.
- Have no reactive functional groups, or only reactive functional groups specifically allowed based on OPPT's risk assessment experience (e.g., blocked isocyanates), or a reactive functional group equivalent weight no less than a defined threshold (e.g., for pendant methacrylates, the equivalent weight threshold is 5,000).
- For polymers with $MW_n \geqslant 10,000$, be incapable of absorbing their weight of water.

As the majority of polymers with aquatic uses (e.g., polymeric flocculants used in water treatment) are cationic in character, they are not eligible for exemption under current polymer exemption criteria.

Environmental Assessment: Exposure

For more than 99% of new chemicals, the focus of OPPT environmental assessment (i.e., ecological assessment) under TSCA is on the aquatic environment. Contamination of surface water is also possible for landfilled chemicals having certain properties (e.g., high water solubility and low biodegradability) if the landfill is in hydrologic contact with groundwater. However, for polymers it is generally assumed that releases to landfills and deep well injection do not result in significant aquatic or terrestrial exposures.

Ecological risk assessment under TSCA is accomplished by using the quotient method (Nabholz, 1991; Nabholz et al., 1993a; Rodier and Mauriello, 1993). This method compares a concern concentration (CC) to an actual or predicted environmental concentration (PEC). Ecotoxicity data are usually expressed as effective concentrations (EC) of the substance (e.g., LC_{50}, EC_{50}, etc.), and CC values are calculated from ECs by the application of assessment factors (AsF). OPPT use of AsF values is outlined in more detail below. PECs are determined from the amount of chemical released, the extent of removal (if any) in wastewater treatment, the extent of dilution of the wastewater by the receiving environment, and the fate of the substance post-release.

The proper starting point for any realistic assessment of environmental exposure is a sound understanding of the substance's physical-chemical properties. Further, depending on the specific substance, release to the environment may result from the manufacture, processing, or industrial use of the substance, from commercial use, and/or from consumer use of household products disposed of down the drain. Except for release from consumer uses, available release data are summarized in an Engineering Report prepared prior to the environmental exposure assessment. Releases are expressed in terms of kilograms (kg) of the chemical substance released per day over a specified number of days per year, for each specific site of release if the sites are known and for generic sites otherwise. Release data transmitted with the Engineering Report do not include consideration of wastewater treatment either on-site or in publicly owned wastewater treatment works (POTWs). Treatment is factored into the assessment in a subsequent step.

To fill data gaps, OPPT may use the U.S. EPA Estimation Programs Interface (EPI) to access computerized estimation methods for melting point, boiling point, and vapor pressure (MPBPVP), water solubility (WSKOW; Meylan et al., 1996), octanol–water partition coefficient (LOG-KOW; Meylan and Howard, 1995), Henry's law constant (HENRY; Meylan and Howard, 1991), and soil/sediment sorption coefficient (PCKOC; Meylan et al., 1992). Also used to predict the octanol–water partition coefficient is the computer program CLOGP (Leo and Weininger, 1985). These methods are based on fragment contribution and fragment-based correction factor approaches, and provide state-of-the-art estimates for most class 1 substances.

For both class 1 and class 2 substances, it is generally assumed by OPPT that water releases are subjected to primary treatment by gravitational settling and activated sludge secondary treatment, at a minimum. If additional or specialized treatment processes are employed, this information is also considered in the assessment if sufficient data are provided. A frequent shortcoming of PMN submissions is that they contain detailed information on the treatment system but only general information on efficiency. For example, removal of biochemical oxygen demand (BOD) in on-site treatment may demonstrably exceed 98% in order to satisfy water permit requirements, but usually there is no way to determine actual removal of the PMN substance itself from this information. Whatever the characteristics of the wastewater treatment system, percent efficiency of removal nearly always must be estimated for new chemicals. For class 1 substances, a fugacity-based multimedia model (Clarke et al., 1995) is often used to provide removal estimates that integrate the key removal processes of sorption, air stripping, and biodegradation. However, the model is not useful, nor is it needed for most polymeric substances, as sorption and/or precipitation are typically the only significant removal process(es).

The final step in the exposure assessment process is to calculate a PEC based on post-treatment release of the PMN substance. In initial review, if discharge is to a water body with unidirectional turbulent flow (i.e., a river or a stream), surface water concentrations usually are calculated by using a simple dilution model, and fate is considered qualitatively. More complex aquatic ecosystem models, such as EXAMS II (U.S. EPA, 1985), in which site-specific ecosystem characteristics as well as transport and transformation parameters can be quantitatively incorporated, are employed only in standard review (Figure 10.2 and U.S. EPA 1994a). If the location of the release site is known, the stream flow for the specific river to which the facility discharges is used. PECs are calculated for mean and low flow conditions using U.S. Geological Survey (USGS) data. If the location is not known, or if multiple release sites are possible, mean and low flows from the stream flow distribution for industrial facilities in the most relevant Standard Industrial Classification (SIC) codes are used (50th and 10th percentile flows for any distribution). If this analysis results in PECs that exceed the

CC, the number of days of exceedance per year is estimated by using the Probabilistic Dilution Model (PDM; U.S. EPA, 1988). In addition to the CC, the model requires input data on kilograms released per day, number of days of release per year, and either location of specific site(s) or relevant SIC code. PDM performs a simple dilution calculation (i.e., does not account for transport or transformation processes after release), but does account for natural variability in stream and effluent flow by incorporating a probability distribution (i.e., flows are assumed to be log-normally distributed) into the calculations.

Environmental Assessment: Toxicity

The purpose of ecotoxicity assessment is (1) to identify as many of the potential effects of a chemical substance toward organisms in the environment as possible and (2) to predict the potency of each effect. As noted above, ecotoxicity data are usually expressed as EC values. For new chemicals, a minimum complete aquatic toxicity profile contains six endpoints with effective concentrations based on 100% active ingredients (a.i.):

- Fish acute value (96-hr LC_{50}).
- Aquatic invertebrate (usually *Daphnia magna*) acute value (48-hr LC_{50}).
- Green algal toxicity value (96-hr EC_{50}).
- Fish chronic value (ChV) from an early life stage toxicity test.
- Aquatic invertebrate ChV from the daphnid partial life-cycle toxicity test.
- Green algal ChV value from the 96-hr toxicity test.

Depending on available data, aquatic toxicity ChV values are EC_{10} estimates that were extrapolated from the Lowest Observed Effect Concentration (LOEC), the No Observed Effect Concentration (NOEC) alone, or the geometric mean of the LOEC and the NOEC. Because OPPT considers that the 96-hr algal toxicity test constitutes a test for chronic toxicity, the geometric mean of the LOEC and the NOEC is often considered to be the algal ChV.

Measured values are preferred by OPPT, but because less than 4.8% of PMNs contain submitted ecotoxicity data (Nabholz et al., 1993a), ECs must be predicted for the great majority of substances (Wagner et al., 1995). Over 95% of the toxicity profiles for new chemicals under TSCA consist entirely of predicted ECs. When ecotoxicity data are submitted with PMNs, they are usually based on acute toxicity with fish. The need to predict the toxicity of chemicals from chemical structure and properties has led to the development of a variety of methods collectively referred to as structure–activity relationships (SARs). The SARs used by OPPT are quantitative in that they are used to predict the potency of a chemical. SARs are developed by OPPT by two general methods: nearest analog analysis and linear regression.

In recent analog analysis, a chemical is compared to one or more analogs. The analogs will have measured toxicity values (e.g., fish LC_{50}). The predicted toxicity value is obtained either by interpolation between analogs (i.e., geometric mean) or by extrapolation from one or more analogs. SAR analysis by linear regression generally involves correlation of chemical structure (or physical properties such as polymer charge density as surrogates for structure) with measured toxicity values.

Chemicals in an SAR are generally homologous chemicals from the same chemical class. SARs for ecotoxicity now exist (Clements, 1988; Clements and Nabholz, 1994; Clements et al., 1993) for many classes of substances subject to TSCA review, and SARs have been immensely useful in assessing new chemicals. Most SARs for ecotoxicity have been developed from data on class 1 substances; SARs for class 2 substances were more difficult to develop. Validation of these SARs is an ongoing process within OPPT (Nabholz et al., 1993b; U.S. EPA, 1994b). SARs are available for all major classes of polymers (see below, "Environmental Concerns for Polymers").

Determination of a CC is the last step in the assessment process for aquatic toxicity. A CC is that concentration of the substance that, if exceeded in the environment, may cause a significant risk of harm. If the exposure assessment does not yield PECs that exceeded the CC, it is assumed that the probability of significant environmental risk from exposure to the substance is too low to warrant direct regulatory action, for example, to limit release. The CC for aquatic life is comparable to the Organisation for Economic Co-operation and Development (OECD) and the European Union (E.U.) Predicted No Effect Concentration (PNEC), and it is analogous to a reference dose (RfD) in risk assessment for human health (Barnes and Dourson, 1988). The CC is expressed in the same units as used for PECs (i.e., milligrams or micrograms per liter). As indicated above, the CC is calculated from ECs in the toxicity profile by the application of assessment factors (AsFs). AsF values are homologous to uncertainty factors (HCN, 1989).

In general, AsFs are used to adjust ECs downward to account for the amount and the quality of data in the toxicity profile and their relevance to predict chronic toxicity under actual environmental conditions. For example, an AsF of 1,000 is used when the toxicity profile contains only one acute toxicity value. This EC is divided by 1,000 to derive the CC. If data are available from several acute but no chronic toxicity tests, a factor of 100 is used; and a factor of 10 is used when the ChV is known from laboratory test data for the presumed most sensitive species, but field data are not available to confirm the ChV. The CC is set equal to the ChV (i.e., the AsF is 1) only when the ChV is derived from field data or, in some cases, microcosms and/or mesocosms. The OPPT database of stream flows is linked with an endangered species database. If the exposure assessment identifies endangered species that exist in the area where release is expected, then the CC is adjusted downward by another factor of 10 for comparison to the PEC.

TESTING

Ecotoxicity

Reliable test data are the foundation of risk assessment and should always be considered preferable to predictions based on SAR and other estimation methods. In ecotoxicity assessment, the issue of which tests should constitute a minimum data set (base set) surfaced at an early stage in the implementation of TSCA (Zeeman and Gilford, 1993). For aquatic exposures, the base set of tests consists of:

- Fish acute toxicity test.
- Aquatic invertebrate acute toxicity test.
- Green algal toxicity test.

For terrestrial exposures, the base set consists of a rodent acute oral toxicity test, a plant early seedling growth test, an earthworm acute toxicity test, a soil microbial community toxicity test, and an avian acute oral toxicity test. The aquatic base set is frequently invoked under TSCA as "tier-one" testing for both new and existing chemicals, and some or all of the tests may be mandated, depending upon the specific data requirements for the substance under review. Higher tier tests consist of subchronic and chronic toxicity tests, toxicity tests using contaminated sediments, microcosms, and field studies (Smrchek et al., 1993).

OPPT's aquatic toxicity test guidelines as well as those of other groups such as the OECD, the E.U., and the American Society for Testing and Materials (ASTM) are generally designed to apply to class 1 substances. These guidelines typically use "clean" dilution water with low levels of dissolved organic carbon (DOC) and total suspended solids (TSS) in fish and invertebrate tests, and low water hardness in algal toxicity tests. However, studies (Cary et al., 1987; Nabholz et al., 1993a) have shown that the toxicity of certain types of polymers may be affected by water hardness or DOC concentration. Toxicity testing for polymers has been modified to address the effects of DOC, hardness, and other conditions encountered in the natural environment. This is discussed more thoroughly below (see "Environmental Concerns for Polymers").

Environmental Fate

The environmental fate of chemical substances is highly variable and ultimately reflects the controlling influence of molecular structure. Although it is true that among the many physical properties of chemicals that might be measured some are more important than others, the most important parameter(s) to measure is not necessarily the same for all substances. From an environmental perspective, vapor pressure may be the most critical property for one substance, whereas for another it may be the substance's water solubility. Transformation processes (e.g., biodegradation and hydrolysis) are even more varied, and in general are also substantially affected by environmental conditions. For these reasons, it has not been possible to develop a fixed yet still affordable set of tests for physical-chemical properties and environmental fate under TSCA.

Fate testing to obtain data on key properties and transformation processes is sometimes required in risk-based 5(e) Consent Orders. In such cases, the nature of the testing generally depends upon the exposed populations and routes of exposure and the outcome of required testing is often used as a trigger for toxicity testing. For example, if a high concern for aquatic toxicity suggests a need for base set testing to resolve uncertainty regarding ecotoxicity, the latter nevertheless may be triggered only if the PMN substance is not readily biodegradable as determined by an appropriate OECD biodegradation test. As another example, if a compound's water solubility or vapor pressure strongly influences the likelihood of exposure via some specific route, health or environmental effects testing may be made contingent upon determination of relevant properties.

A fate testing strategy designed to account for the complexity of environmental transport and transformation phenomena was developed for application in exposure-based PMN review (XBR). When polymers are expected to be released to water in significant volumes, it is likely that the XBR fate testing strategy will apply to them. The strategy borrows some aspects of the base set approach in ecotoxicity testing but is also intended to be flexible. Aspects relevant to aquatic exposures are outlined in Table 10.2. An important feature is that U.S. EPA generally limits required fate testing to one or two tests for substances with estimated production volumes $\geqslant 100,000$ kg/year (i.e., 220,000 lb/year), whereas if the volume is $\geqslant 454,000$ kg/year (i.e., 1,000,000 lb/year) three or more tests may be required if the data are deemed essential.

Table 10.2. Fate testing strategy for aquatic exposures in exposure-based review

Exposure criterion	Recommended tests
Production volume ≥ 100,000 kg/yr and at least one other XBR criterion[a]	Wastewater treatment removal: Semi-continuous Activated Sludge (SCAS), Porous Pot, or Activated Sludge Sorption Isotherm
	-and-
	One of the following: water solubility, vapor pressure, soil/sediment sorption isotherm, ready biodegradability, or hydrolysis as appropriate
Production volume ≥ 454,000 kg/yr (1,000,000 lb/yr) and at least one other XBR criterion[a]	Wastewater treatment removal: tests listed above
	-and-
	At least two of the following: water solubility, vapor pressure, soil/sediment sorption isotherm, ready biodegradability, anaerobic biodegradability, or hydrolysis as appropriate

[a]General population: exposure via surface drinking water of ≥ 70 mg/yr, and environment: release to surface water of ≥ 1,000 kg/yr.

Existing OPPT test guidelines in the areas that the XBR fate strategy addresses are primarily intended to apply to class 1 substances, not polymers. Nevertheless, the strategy is interpreted broadly by OPPT, and acceptable test protocols do not need to be based strictly on existing published guidelines. For example, data from an appropriate jar test (ASTM, 1994) or other laboratory test designed to measure the performance of sedimentation aids in wastewater clarification (Halverson and Panzer, 1980) may be more relevant than guideline tests for determining the fate of a polymeric substance in wastewater treatment. Particularly for polymers, it is prudent for the chemical manufacturer or importer to submit the protocol to OPPT for review and approval before testing begins.

FATE ASSESSMENT WITH POLYMERS

For convenience, polymers may be pragmatically divided into four classes based on charge: anionic (negative charge), cationic (positive charge), amphoteric (positive and negative charges present on the same polymer), and nonionic (neutral charge). Electronic charge, MW, and solubility/ dispersibility in water are typically the most important properties of polymers for fate assessment. Charge density (i.e., number of charges per unit of MW) may also be important but probably plays a greater role in toxicity assessment than in fate assessment (see below, "Environmental Concerns for Polymers").

The vast majority of synthetic polymers are essentially nonbiodegradable, a fact that has been known for many years (e.g., Alexander, 1973). Even modified natural polymers such as carboxymethylcellulose and cellulose acetate butyrate, which have an appreciable degree of substitution, are not significantly biodegradable. Some modified natural polymers and synthetic polymers with ester linkages or other labile groups incorporated into the main chain of the polymer are biodegraded under favorable conditions. Examples include poly(β-hydroxybutyric acid), a naturally occurring bacterial energy storage product, poly(ε-caprolactone), poly(glyoxylic acid), and poly(tetramethylene succinate) (Bailey and Gapud, 1986; Park et al., 1989; Pranamuda et al., 1995). However, rates of biodegradation that may be significant for terrestrial environments (e.g., landfills) may not be so for wastewater treatment systems. Wastewater retention times are much shorter than typical residence times in terrestrial environments. Thus, even so-called biodegradable polymers are not necessarily significantly biodegraded in sewage treatment.

All PMN substances, including polymers, are evaluated for their biodegradability based on molecular structure if submitted data are insufficient. In practice, the few biodegradable polymers among the many recalcitrant ones are easily identified because a polymer's biodegradability is usually a major reason for the submitter's desire to market the substance. In these cases, the PMN generally contains experimental data on biodegradability, and this characteristic figures prominently in a Pollution Prevention (PP) claim.

As air stripping obviously is not a major loss mechanism for polymers, assessment of the removal of polymers in wastewater treatment essentially reduces to an assessment of removal by sorption and/or precipitation. In the review of new chemicals under TSCA, the treatability of a polymer discharged to wastewater treatment is inferred from the polymer's charge, MW, and solubility/dispersibility unless actual data accompany the PMN (rarely the case). It is generally assumed that the sequence of treatment processes is limited to primary clarification followed by activated sludge secondary treatment, which is typical for POTWs. Nonionic, cationic, and amphoteric polymers with $MW_n > 1,000$ are assumed to partition mainly to the solids phase and to be 90% removed relative to the total influent concentration. The 90% figure was selected because it represents a typical level of solids removal in POTWs (U.S. EPA, 1982). The remaining fraction (10%) is assumed to be discharged to receiving waters although of course it is likely that this material is in the form of polymer sorbed to sludge solids. For a polymer of $500 < MW_n < 1,000$, a lower removal rate (typically 50–90%) may be assumed, depending on the polymer's structure and properties.

Anionic polymers with negligible water solubility and dispersibility are assumed to behave similarly to nonionic polymers, but lower removal rates are assumed for polyanionics having appreciable solubility or dispersibility. Percent removal varies with MW and polymer type, but default values are typically assigned by OPPT as follows:

$$MW_n < 5,000 \qquad\qquad 0 \text{ to } 50\%$$

$$5,000 \leqslant MW_n < 20,000 \qquad 50\%$$

$$20,000 \leqslant MW_n < 50,000 \qquad 75\%$$

$$MW_n \geqslant 50,000 \qquad\qquad 90\%$$

This scheme is consistent with data for elimination of poly(carboxylic acids) in the Semi-continuous Activated Sludge (SCAS) and OECD Confirmatory tests (Table 10.3; from Opgenorth, 1992), and it assumes that the removal of polyanionics will be lower than for cationic or neutral polymers owing to the net negative charge of microbial cell surfaces. Polycarboxylates are

Table 10.3. Biodegradation and elimination of polycarboxylates in tests with activated sludge[a]

Polycarboxylate[b]	Ultimate biodegradation (ThCO₂), %[c]	SCAS test (DOC loss), %[d]	Confirmatory test (removal), %
P(AA) 1,000	45	45	—[e]
P(AA) 2,000	20	21	—
P(AA) 4,500	9	40	27
P(AA) 10,000	16	58	—
P(AA) 60,000	—	93	—
P(AA-MA) 12,000	31	83	70
P(AA-MA) 70,000	20	95	82

[a]*Source:* Opgenorth, 1992; see also Chapter 6.

[b]P = poly; AA = acrylic acid; MA = maleic acid. MW_n given are approximate.

[c]Percent of theoretical CO_2 (ThCO_2) formation after contact with activated sludge for 30–90 days.

[d]SCAS = Semi-continuous Activated Sludge; DOC = dissolved organic carbon.

[e]— = not reported in Opgenorth (1992).

widely used in detergents and other cleaning products and are generally either homopolymers of acrylic acid or copolymers of acrylic and maleic acids. Their environmental properties and toxicity have been extensively reviewed by Opgenorth and others (see Chapter 6).

As a general rule, for purposes of estimating environmental concentrations, polymers discharged from wastewater treatment are typically assumed not to undergo further removal by in-stream fate processes. Precipitation of polyanionics by alkali metal cations (e.g., Ca^{2+} and Mg^{2+}), polymer binding to DOC, and other sorption phenomena control the bioavailability of polymers (Cary et al., 1987; Nabholz et al., 1993a). Factors that affect bioavailability are considered in setting the CC for a polymer; key factors are addressed below.

ENVIRONMENTAL CONCERNS FOR POLYMERS

General Aspects

In toxicity assessments with polymers, a distinction is made between polymers with minimal low molecular weight (LMW) material (i.e., $MW_n > 1,000$, with $<25\% < 1,000$ and $<10\% < 500$) and polymers with either $MW_n < 1,000$ or polymers with significant amounts of LMW material (i.e., $\geqslant 25\% < 1,000$ and $\geqslant 10\% < 500$) (U.S. EPA, 1996). Water-soluble or dispersible polymers with either $MW_n < 1,000$ or significant amounts of LMW material are of concern because of their toxicologic similarity to polymers with $MW_n > 1,000$, but they may also be absorbed through biological membranes and cause systemic effects. Ecotoxicity assessments for these polymers are generally based on both the type of polymer and the type of functional group(s) in the LMW components (Nabholz et al., 1993a). Polymers with $MW_n > 1,000$ are not absorbed through the respiratory membranes of aquatic organisms, and thus toxicity is manifested either through direct surface-active effects on outer membranes of aquatic organisms or indirectly via chelation of essential nutrients, or both ways (Nabholz et al., 1993a). All polymers are assessed as polymers, but, in addition, polymers with $MW_n < 1,000$ and polymers with significant amounts of LMW oligomers are also assessed as monomers.

Insoluble polymers are not expected to be toxic unless in the form of finely divided particles. The toxicity of insoluble particles does not depend upon the chemical structure of the polymer and results from occlusion of respiratory organs such as gills. In this case, aquatic toxicity occurs only at high concentrations (Wagner et al., 1995); acute toxicity values are generally $> 100\,mg/L$ and chronic values $> 10\,mg/L$.

Anionic Polymers

Polyanionics with $MW_n > 1,000$ that are water-soluble or dispersible in water are of concern for aquatic toxicity from either direct or indirect toxicity. Anionic polymers are divided into two subclasses for assessment purposes: poly(aromatic acids) and poly(aliphatic acids).

Poly(Aromatic Sulfonate/Carboxylate): Toxicity

Poly(aromatic sulfonate/carboxylate) polymers with $MW_n > 1,000$ are of moderate concern for toxicity to aquatic organisms (i.e., LC_{50} or EC_{50} values between $1\,mg/L$ and $100\,mg/L$) (Wagner et al., 1995), depending on the monomers they contain (Table 10.4; polymers designated 1 to 15). Dominant monomers associated with polymer toxicity thus far identified are: carboxylated diphenolsulfones, sulfonated diphenolsulfones, sulfonated phenols, sulfonated cresols, sulfonated diphenylsulfones, and sulfonated diphenylethers. Monomers of low concern are sulfonated naphthalene and sulfonated benzene.

The strongest evidence for direct aquatic toxicity of poly(aromatic sulfonate/carboxylate) polymers is based on data for a carboxylated bi-phenolsulfone polymer (polymer 1). This polymer had a MW_n clearly $> 1,000$ and negligible percentages of LMW oligomers. The green algal toxicity test was repeated three times with growth/test medium at increasing hardness to determine if the observed toxicity was due to the indirect effect of overchelation of nutrient divalent elements. There was no reduction in toxicity with increased hardness (i.e., algal 96-hr EC_{50} values of 24, 20, and $47\,mg/L$, respectively, for hardness levels of 46, 152, and $160\,mg/L$ as $CaCO_3$). The mechanism of toxicity is unknown but this polymer was algicidal..

Base set toxicity values for other relatively toxic polymers (polymers 2 to 13) range to a low value of $2\,mg/L$ (polymer 8). The toxicity of these polymers is undoubtedly a combination of direct toxicity from the polymers and systemic toxicity from LMW oligomers, which are capable of being absorbed through respiratory membranes.

The nearest analog SAR method is used by OPPT to predict the ecotoxicity of poly(aromatic acids). For example, although there are no data for polymers with substituted phosphoric acid monomers, toxicity of poly(aromatic phosphate) polymers is assumed by OPPT to be equivalent to poly(aromatic sulfonates/carboxylates) until test data can be obtained. It

Table 10.4. Aquatic toxicity data for polyanionics: poly(aromatic sulfonates) and poly(aromatic carboxylates)[a,b]

Polymer no.[c]	MW_n/ < 1,000/ < 500	F96	D48	GA96	GAChV
				—mg/L—	
		Biphenolsulfone–COOH[b]			
1	12K/0/0	72.0	86.0	40.0	< 12.5
1	12K/0/0	—	—	24.0	7.0
1	12K/0/0	—	—	47.0	18.0
1	12K/0/0	—	—	20.0	< 12.5
		Biphenolsulfone–SO$_3$H			
2	> 1K/40/15	5.0	145	20.0	—
3	1K/—/15	5.1	189	44.0	< 7.5
4	> 0.4K/100/10	9.7	136	9.3	—
5	1K/—/15	23.0	950	12.0	6.2
		Biphenolsulfone–SO$_3$H and cresol–SO$_3$H			
6	1.1K/0/0	3.2	—	—	—
		Biphenolsulfone–SO$_3$H and biphenyl ether–SO$_3$H			
7	0.6K/95/40	80.0	120	—	—
7	0.6K/95/40	32.0	—	—	—
		Phenol–SO$_3$H			
8	1K/—/—	2.0	—	—	—
9	> 1K/19/3	510	—	—	—
		Cresol–SO$_3$H			
10	12K/—/0	3.0	—	—	—
11	1K/—/—	⩽ 2.3	—	—	—
		Biphenylsulfone–SO$_3$H			
12	1.1K/—/—	3.0	—	—	—

Continued on next page

Table 10.4.   Continued

Polymer no.[c]	MW_n/<1,000/<500	F96	D48	GA96	GAChV
				mg/L	
	Biphenyl ether–SO_3H				
13	0.6K/90/40	30.0	120	—	—
	Naphthalene – SO_3H				
14	8K/—/—	150	>500	340	44.2
	Benzene – SO_3H				
15	3K/30/1	600	900	800	—

[a]Polymer no. = polymer identification number in text; MW_n/<1,000/<500 = average-number molecular weight of polymer in thousands (K) with % less than 1,000 and % less than 500; effective concentration (EC) for F96 = fish 96-hr LC_{50}, D48 = daphnid 48-hr LC_{50}, GA96 = green algal 96-hr EC_{50}, and GAChV = green algal Chronic Value. All ECs are based on 100% active ingredients or 100% polymer solids and nominal concentrations.

[b]Polymers are grouped by dominant monomer by weight in polymer backbone.

[c]Replicate tests were run for polymers 1 and 7; "—" indicates data not reported.

also appears that the most common chemical moiety for this subclass is phenol. Thus, other polymers differing only in the type of phenolic moiety should be assumed to have toxicity comparable to that of poly(aromatic sulfonates/carboxylates).

Poly(Aliphatic Acids): Toxicity

Poly(aliphatic acid) polymers are members of a subclass of polyanionics, which often contain one or a combination of carboxylic acid, phosphinic acid, and sulfonic acid monomers. At pH 7, these polymers are of concern only for their indirect moderate toxicity to green algae. Toxicity toward fish and aquatic invertebrates (Table 10.5) is consistently low, with LC_{50} values >100 mg/L (Wagner et al., 1995). For example, the mean LC_{50} for polyanionics in Table 10.5 toward fish and aquatic invertebrates is >225 mg/L. Measured acute toxicity values for other polyanionics submitted to the NCP under Section 5 of TSCA had an average fish $LC_{50} > 780$ mg/L ($n = 43$) and an average daphnid $LC_{50} > 560$ mg/L ($n = 22$), at neutral pH. Results

Table 10.5. Aquatic toxicity data for polyanionics: poly(aliphatic acids)[a,b]

Polymer no.	MW_n/ <1,000/ <500	F96	D48	GA96	GAChV
				mg/L	
Distance between acids: 0 carbons: $[C(A)C(A)]_x$					
16	0.5K/—/—	1,140	>339	560	280
16	0.5K/—/—	2,500	—	—	—
Distance between acids: 1 carbon: $[C(A)C]_x$					
17	3K/—/—	330	>440	7.44	—
18	3.5K/—/—	>500	>500	3.13	—
19	2.5K/5/3	>225	—	37.4	24.0
20	0.9K/49/19	—	>1,000	11.0	4.7
21	10K/35/5	—	—	7.6	0.50
22	1.4K/41/8	—	—	5.5	0.54
Distance between acids: 1.5 carbons: $[C(A)C(A)C]_x$					
23	0.6K/88/69	—	500	66.0	—
Distance between acids: 2 carbons: $[C(A)CCC(A)]_x$					
24	3K/0/0	—	—	150	93.0
Distance between acids: 1 carbon; distance from backbone: 3 carbons: $[C(A)C]_x + [C(CCCA)C]_x$					
25	1K/15/5	>1,000	1,800	57.0	36.0
Distance between acids: 1 carbon: chelating monomer diluted 3 to 1 by nonchelating monomer: $[C(A)C]_3 + [C(CONH_2)C]_1$					
26	>1K/10/1	>40.0	>40.0	>500	>100

[a]Polymer no. = polymer identification number in text; MW_n/1,000/ <500 = average-number molecular weight of polymer in thousands (K) with % less than 1,000 and % less than 500; effective concentration (EC) for F96 = fish 96-hr LC_{50}, D48 = daphnid 48-hr LC_{50}, GA96 = green algal 96-hr EC_{50}, and GAChV = green algal Chronic Value. All ECs are based on 100% active ingredients or 100% polymer solids and nominal concentrations, and standard algal growth media with hardness between 10 and 24 mg/L as $CaCO_3$.

[b]A = acid functionality monomers (e.g., carboxylic acid, phosphinic acid, and sulfonic acid). Polymers are grouped by distance between acids, which is measured by counting the average number of polymer backbone carbons between acids.

[c]Replicate tests were run for polymer 16; "—" indicates data not reported.

were reported in terms of 100% polymer active ingredients (a.i) or 100% polymer solids.

The moderate toxicity of anionic polyaliphatics to algae appears to be associated with polymer chelation of nutrient elements needed by algae for growth (Nabholz et al., 1993a). The elements include calcium, magnesium, and probably iron. Available test data further suggest that polymer potency with algae is directly related to carbon distances between acids. The distance between acids can control the strength of chelates formed between two acids and polyvalent cations. Homopolymers of acrylic acid (that is, poly(acrylic acids)) that contain carboxylic acid on every other carbon in the polymer backbone have been the most toxic of poly(aliphatic acids) to green algae (Table 10.5; polymers 17 to 22). Polymers 17 to 22 present a geometric mean 96-hr EC_{50} value with algae of 8.6 mg/L. This geometric mean is used below to compare the aquatic toxicity of poly(acrylic acids) to other polyanionics.

In contrast, if acids are moved closer together or farther apart along the polymer backbone, algal toxicity decreases. Test data for homopolymers of maleic acid (polymer 16), which have carboxylate acid groups on every carbon of the polymer backbone, indicate low toxicity to algae (96-hr EC_{50} = 560 mg/L), 65 times less toxicity relative to poly(acrylic acid), and a relatively weak capacity to chelate polyvalent cations. Test data for polymers that have acids farther apart than 1 carbon (i.e., 1.5 carbon separation (polymer 23) and 2 carbon separation (polymer 24) showed 8 times and 17 times less toxicity, respectively, when compared to polymers 17 to 22.

Toxicity of polyanionics to algae may also be reduced by (1) adjusting the distance between acids by moving some acids farther from (i.e., pendant to) the polymer backbone and/or (2) randomly "diluting" the effect of monomers associated with chelation of cations (e.g., acrylic acid) with monomers that do not chelate (e.g., acrylamide). For example, in polymer 25 (Table 10.5) acrylic acid has been randomly reacted with a second monomer that has a carboxylic acid on every other carbon. The second monomer's carboxylic acids are pendant to the backbone by a distance of 3 carbons. The green algal 96-hr EC_{50} of polymer 25 is 57 mg/L; so it is 7 times less toxic than poly(acrylic acids). Test data for polymer 26 show the effect on toxicity of an added nonchelating (or weakly chelating) monomer. The green algal 96-hr EC_{50} of polymer 26 was > 500 mg/L, a toxicity reduction of over 60 times compared to the green algal 96-hr EC_{50} of poly(acrylic acids).

Based on the above information, poly(aliphatic acids) have been grouped in Table 10.5 by the average distance between the acids on the polymer. The toxicity of these polymers can be predicted by using the nearest analog SAR method and knowing the monomer composition of a polymer and the reaction sequence (i.e., random reaction of monomers or blocked reactions). If monomers are blocked in order to use the polymer as a surfactant or a dispersant, then the polymer could be toxic to aquatic organisms by a surface-related mechanism of action.

Poly(Aliphatic Acids): Mitigation of Toxicity

Available information suggests that moderate toxicity to algae from poly(aliphatic acids) will be observed only in soft water that has a hardness of less than 30 mg/L as $CaCO_3$. The hardness of algal growth media recommended in standard test guidelines (e.g., OECD) is approximately 15 to 24 mg/L. Standard algal toxicity test guidelines generally seek to maximize algal growth and to measure "intrinsic" (i.e., baseline) toxicity and thus recommend media with low hardness. However, toxicity measured in the standard green algal toxicity tests is potentially an overestimate of toxicity for most natural surface waters.

If Ca^{2+} salts of poly(aliphatic acids) are tested in standard algal media, or if the same polymers are tested in media with a hardness of near 150 mg/L $CaCO_3$, algal toxicity is mitigated (Table 10.6). Many poly(aliphatic acids) are used as scale inhibitors in industrial and commercial applications and are released to the environment already chelated with Ca^{2+} and/or Mg^{2+}. Other major uses such as in detergents and cleaning products should lead to release of the polymers to wastewater treatment, where chelation and removal by sorption occur. For polymers released to sewage effluent, mitigation of toxicity may also occur from divalent cations in receiving waters; the average hardness of freshwater in the United States is about 120 mg/L as $CaCO_3$ (Nalco, 1988).

OPPT began to recommend toxicity mitigation testing for poly(aliphatic acids) in 1989. PMN test data for polymers 19, 20, and 22 demonstrated mitigation factors of from 14 to 380 times as the amount of Ca^{2+} in the testing environment increased. Test data for polymers 23 and 25 indicated less mitigation, with mitigation factors of 12 and 8.9, respectively. However, mitigated EC_{50} values of polymers 23 and 25 were similar to the mitigated EC_{50} values of polymers of 19, 20, and 22. The mitigated 96-hr EC_{50} values were 500 to 950 mg/L for polymers 19, 20, 22, 23, and 25.

Currently recommended algal toxicity testing for poly(aliphatic acids) under Section 5 of TSCA generally consists of three tests: Test 1 — "neat" polymer (i.e., polymer as is) at pH 7.5 in standard algal growth medium; Test 2 — polymer with stoichiometrically equivalent Ca^{2+} added to the polymer stock solution at pH 7.5 and tested in standard growth medium; and Test 3 — neat polymer at pH 7.5 tested in modified algal growth medium. Modified growth medium contains calcium alone or calcium and magnesium combined to attain a hardness of about 150 mg/L. If Ca^{2+} and Mg^{2+} are added together, they should be present in a 2:1 (Ca:Mg) ratio.

Table 10.6. Mitigation of toxicity toward green algae from polyanionics by calcium ions $(Ca^{2+})^a$

Polymer no.	$MW_n / <1,000 / <500$	Effect type	EC (mg/L)	Hardness[b] or eq. Ca^{2+}	MF	Mean MF
Distance between acids: 1 carbon: $[C(A)C]_x$						
19	2.5K/5/3	GA96	37.4	H 15		
19	2.5K/5/3	GA96	500	$1.0\,Ca^{2+}$	13	
19	2.5K/5/3	GAChV	24.0	H 15		
19	2.5K/5/3	GAChV	350	$1.0\,Ca^{2+}$	15	14
20	0.9K/49/19	GA96	11.0	H 15		
20	0.9K/49/19	GA96	600	H 315	55	
20	0.9K/49/19	GAChV	4.7	H 15		
20	0.9K/49/19	GAChV	100	H315	21	34
22	1.4K/41/8	GA96	5.5	H 20		
22	1.4K/41/8	GA96	69.0	H 41		
22	1.4K/41/8	GA96	780	H 158	140	
22	1.4K/41/8	GAChV	0.54	H 20		
22	1.4K/41/8	GA96	6.25	H 41		
22	1.4K/41/8	GA96	550	H 158	100	
22	1.4K/41/8	GA96	950	$4.3\,Ca^{2+}$	170	
22	1.4K/41/8	GAChV	600	$4.3\,Ca^{2+}$	1,100	380
Distance between acids: 1.5 carbons: $[C(A)C(A)C]_x$						
23	0.6K/88/69	GA96	66.0	H 14		
23	0.6K/88/69	GA96	800	$1.0\,Ca^{2+}$	12	12
Distance between acids: 1 carbon; distance from backbone: 3 carbons: $[C(A)C]_x + [C(CCCA)C]_x$						
25	1K/15/5	GAChV	36.0	H 15		
25	1K/15/5	GAChV	320	H 315	8.9	8.9

[a]Polymer no. = polymer identification number; $MW_n / <1,000 / <500$ = average-number molecular weight of polymer in thousands (K) with percent less than 1,000 and percent less than 500; Effect type: GA96 = green algal 96-hr EC_{50} or GAChV = green algal chronic value; EC = effective concentration; hardness = mg/L as $CaCO_3$ or equivalents of Ca^{2+} added to polymer; MF = mitigation factor; and mean MF = average MF for a polymer. All ECs are based on 100% active ingredients or 100% polymer solids and nominal concentrations.

[b]Calcium is added by either increasing the hardness of the algal growth medium or testing the Ca^{2+} salt of the polymer. Data have been grouped by the average distance between acids in terms of polymer backbone carbons.

The purpose of the above scheme is to simulate conditions of possible release. For example, polymers used as scale inhibitors are expected to be released to the environment chelated with Ca^{2+} and Mg^{2+} and therefore should be tested with all three tests. Test 1 measures intrinsic toxicity. If the 96-hr EC_{50} is $> 100\,mg/L$, and the ChV is $> 10\,mg/L$, then no further testing need be done because of low concern for intrinsic toxicity. Test 2 simulates the release of polymer in a divalent salt chelated form (e.g., use as a scale inhibitor). Test 3 simulates the release of the neat polymer from manufacturing and processing. For polymers not expected to be used as scale inhibitors or otherwise released as Ca^{2+} and/or Mg^{2+} salts, Test 2 can be eliminated.

If the site of release is known and the hardness of receiving water is known at the site, then the hardness of the medium can be matched to site-specific conditions. In some cases, PMN submitters have combined Tests 2 and 3 into one test by testing Ca^{2+} salts of polymers in a moderately hard medium. This is acceptable, but one must keep in mind that the green algal 96-hr EC_{50} due to hardness alone is 1,140 mg/L, and the corresponding ChV is 80 mg/L.

It should be noted that mitigation testing must be conducted at near-neutral pH. Mitigation of poly(aliphatic acids) fails when acid polymers are tested nonneutralized. Toxic effects of low pH to algae cannot be overcome by adding Ca^{2+} to the polymer or to the medium.

Nonionic Polymers

Nonionic polymers with $MW_n > 1,000$, $<25\% < 1,000$, and $<10\% < 500$ are generally of low concern for ecotoxicity because they often have negligible water solubility. If a nonionic polymer is water-soluble (or dispersible) and has monomers reacted via random order, then concern for aquatic toxicity is still low with base set LC_{50}/EC_{50} values expected to be $> 100\,mg/L$. However, as noted above, if monomers are blocked to use the polymer as a surfactant or a dispersant, such blocked polymers could be toxic to aquatic organisms by a surface-active mechanism of action and should be tested. Nonionic polymers with significant amounts of oligomer content (i.e., $>25\% < 1,000$ and $>10\% < 500$) may be of concern on the basis of the bioavailability and the toxicity of the LMW material, and the LMW oligomers are assessed as monomers. Nonionic polymers with $MW_n < 1,000$ are assessed as monomers.

Cationic Polymers

Cationic polymers of concern for aquatic toxicity include polymers that contain a net positively charged atom or that contain groups that can reasonably be anticipated to become cationic in water (U.S. EPA, 1996). Atoms with a net positive charge include, but are limited to, nitrogen as quaternary ammonium, phosphorus as phosphonium, and sulfur as sulfonium. Groups anticipated to become cationic in water include, but are not limited to, aliphatic primary, secondary, and tertiary amines. Forms of nitrogen not currently included for concern for ecotoxicity are (a) aromatic nitrogens, unless they are quaternarized, (b) nitrogens directly substituted to benzene (as in aniline), (c) amides, (d) nitriles, and (e) nitro groups. As with polyanionics, polycationics of concern have to be either water-soluble or dispersible in water. Cationic polymers that are solids and are only to be used in the solid phase are of low concern; specifically, dispersed beads of polycationics are of low concern.

Polycationics are assessed by OPPT according to their type of polymer backbone, whether carbon-based, silicone-based (e.g., Si–O), or natural (e.g., chitin, starch, and tannin). The type of backbone can influence aquatic toxicity and some physical-chemical properties. As shown in Table 10.7, aquatic toxicity data for polycationics are grouped according to polymer backbone.

The aquatic toxicity of polycationics in clean water (i.e., water with a low total organic carbon (TOC) content of $<2\,mg/L$) increases exponentially with higher charge density until toxicity becomes asymptotic. Polycationics have been shown to elicit acute toxic effects in aquatic organisms by physically disrupting respiratory (e.g., gill) membranes, interfering with O_2 exchange (Biesinger and Stokes, 1986). It is assumed that polycationics strongly adsorb to all biological membranes that are anionic. It is also assumed that chronic toxicity occurs by the same mechanism for polymers with minimal amounts of LMW oligomers (i.e., $<25\% < 1,000$ and $<10\% < 500$). For polycationics with $MW_n < 1,000$ and polymers with significant amounts of LMW oligomers (i.e., $>25\% < 1,000$ and $>10\% < 500$), systemic toxicity is also possible (Nabholz et al., 1993a).

Based on existing information, aquatic toxicity of polycationics is most strongly influenced by cationic charge density and type of polymer backbone (Table 10.7), *not* (a) pH dependence (e.g., quaternary ammoniums versus aliphatic primary, secondary, or tertiary amines), (b) backbone position of cations, or (c) MW_n. For example, for polymers with silicone-based backbones (polymers 62 to 70; Table 10.7), the largest difference between daphnid 48-hr LC_{50} values due to charge density is about 2,300 times (i.e., polymer 62 compared to polymer 69). The difference between daphnid 48-hr LC_{50} values with a tertiary amine (polymer 65) and a quaternary ammonium (polymer 66) was only 5 times.

The effect of MW may be greatest for test species with relative high surface to volume ratios (i.e., algae > daphnids > fish). Polymers 28 and 29 have the same charge density of 0.7% amine-N but differ greatly in MW_n

Table 10.7. Aquatic toxicity data for polycationics grouped by type of polymer backbone[a,b]

Polymer no.[c]	A-N (%)	MW_n/<1,000/<500	Cat. pos.	Cat. type	F96	D48	GA96	GAChV
						(mg/L)		
		Carbon-Based Backbone Polymers						
27	0.69	6.2K/0/0	P11	3	4.6	—	1.3	0.16
28	0.70	1.8K/—/—	—	—	9.2	300	2.2	0.88
28	0.70	1.8K/—/—	—	—	8.5	310	—	—
28	0.70	1.8K/—/—	—	—	3.9	—	—	—
29	0.70	8,000K/0/0	P4	4	3.3	—	>360	130
30	0.70	8,000K/0/0	P4	4	10.6	28.2	—	—
31	0.70	8,000K/0/0	P4	4	10.0	40.0	—	—
32	1.1	6K/2/1	B	3	30.0	—	—	—
33	1.4	2.4K/13/2	B	2	17.0	16.0	0.52	0.27
34	2.0	2,500K/—/—	P2	3,4	0.97	1.7	—	—
34	2.0	2,500K/—/—	P2	3,4	2.3	—	—	—
35	2.0	1,100K/0/0	P5	3	0.64	—	—	—
35	2.0	1,100K/0/0	P5	3	1.2	—	—	—
36	2.1	19,000K/0/0	P4	3	0.84	—	—	—
37	3.0	100K/—/—	P2	3	0.90	—	—	—
38	3.1	180K/0/0	P1	4	0.32	—	—	—
39	3.3	5,000K/0/0	P3	2,3	1.5	0.09	0.035	0.006
40	3.4	50K/0/0	P4	4	0.60	—	—	—
41	3.4	50K/0/0	P4	4	0.30	—	—	—
42	3.6	250K/0/0	P4	4	0.73	0.18	0.025	0.013
43	4.5	1,000K/0/0	P4.5	4	0.31	—	—	—
44	4.6	1,000K/0/0	P5	4	0.90	—	—	—
45	5.1	220K/0/0	P4	4	0.76	—	—	—
46	6.0	>5.0K/—/—	B	4	0.15	—	—	—
46	6.0	>5.0K/—/—	B	4	0.16	—	—	—
46	6.0	>5.0K/—/—	B	4	0.29	—	—	—
47	6.4	1K/38/12	B	4	0.72	0.073	0.014	0.006
48	7.8	1K/32/23	B	2	1.0	2.9	0.015	0.007
49	8.0	5.0K/—/—	B	4	0.13	0.34	0.016	—
49	8.0	5.0K/—/—	B	4	0.22	—	—	—
49	8.0	5.0K/—/—	B	4	0.22	—	—	—
50	8.1	1.5K/20/5	B	2,3	0.73	—	0.047	0.03
51	8.1	58K/1/1	P2.5	4	0.18	—	0.02	0.008
52	9.2	4K/49/19	B	3,4	0.50	—	—	—
53	11.0	1.8K/55/26	B	3,4	0.22	0.58	0.07	0.034
54	11.0	1.5K/44/26	B,P1	1,2,3	0.20	—	—	—
55	12.0	8K/—/—	B	2	1.9	1.2	—	—
56	14.0	2K/20/3	B	2	0.10	0.69	0.061	0.03
56	14.0	2K/20/3	B	2	0.16	0.38	0.052	0.01
56	14.0	2K/20/3	B	2	0.10	—	—	—
57	14.0	2K/—/—	B	2	0.30	0.78	0.08	0.044
58	14.0	2K/25/5	B	2	0.072	0.276	0.032	0.017
58	14.0	2K/25/5	B	2	0.084	2.9	0.08	0.050
58	14.0	2K/25/5	B	2	0.16	—	—	—
59	15.0	25K/—/—	B	2	0.26	0.26	—	—
59	15.0	25K/—/—	B	2	0.24	—	—	—

Table 10.7. Continued

Polymer no.[c]	A-N (%)	MW_n/<1,000/<500	Cat. pos.	Cat. type	F96	D48	GA96	GAChV
						(mg/L)		
Carbon-Based Backbone Polymers								
60	17.0	50K/1/1	B	2,3	0.45	—	—	—
61	20.0	25K/—/—	B	2	0.32	0.17	—	—
61	20.0	25K/—/—	B	2	0.32	—	—	—
61	20.0	25K/—/—	B	2	0.32	—	—	—
61	20.0	25K/—/—	B	2	0.32	—	—	—
61	20.0	25K/—/—	B	2	0.23	—	—	—
61	20.0	25K/—/—	B	2	0.20	—	—	—
Silicone-Based Backbone Polymers								
62	0.40	1.1K/23/19	P4.5	2,3	300	370	0.11	0.09
63	0.50	30K/0/0	P3	1	—	44.0	—	—
64	0.70	7.6K/—/—	P5.5	1,2	—	1.0	—	—
65	0.70	7.6K/—/—	P4.5	1,2	—	0.50	—	—
66	0.70	7.8K/—/—	P4.5	4	—	0.10	—	—
67	0.70	7.8K/—/—	P4.5	1,2	6.1	1.0	—	—
67	0.70	7.8K/—/—	P4.5	1,2	—	1.1	—	—
67	0.70	7.8K/—/—	P4.5	1,2	—	1.4	—	—
68	0.78	10K/4/2	P8.5	1,2	28.0	15.0	—	—
69	2.6	1.7K/17/5	P7	4	0.65	0.16	9.6	—
70	5.5	1.1K/5/0	P7	2	1.17	1.04	—	—
Natural-Based Backbone Polymers								
71	0.07	2,300K/0/0	P4	3	>850	—	>1,000	>1,000
72	0.2	30,000K/0/0	P3	4	>1,000	177	—	—
73	0.29	30,000K/0/0	P3.3	3	>1,000	>1,000	—	—
74	0.4	10,000K/0/0	P3	3	>1,000	>1,000	—	—
75	0.93	22,000K/0/0	P5	4	1.18	—	—	—
75	0.93	22,000K/0/0	P5	4	0.37	—	—	—
76	0.93	980K/—/—	P10	4	—	130	—	—
77	1.7	73K/20/10	P4	4	0.86	—	—	—
78	3.4	1.7K/30/25	P4	4	1.5	26.0	0.62	0.39
79	4.3	500K/0/0	B	1	570	11.0	>480	>480

[a]Polymer no. = Polymer identification number; A-N = cationic charge density as % amine nitrogen; MW_n/<1,000/<500 = average-number molecular weight of polymer in thousands (K) with percent less than 1,000 and percent less than 500; Cat. pos. = position of cation in polymer where: B = cation in backbone or P# = pendant cation, with number (#) of carbons from backbone; Cat. type = primary amine (1), secondary amine (2), tertiary amine (3), and quaternary (4); and effective concentration (EC) for F96 = fish 96-hr LC_{50}, D48 = daphnid 48-hr LC_{50}, GA96 = green algal 96-hr EC_{50}, and GAChV = green algal Chronic Value.

[b]All ECs are based on 100% active ingredients or 100% polymer solids, nominal concentrations. TOC of dilution water was less than 2 mg/L; "—" indicates data not reported.

[c]Replicate data are given for polymers 28, 34, 35, 46, 49, 56, 58, 59, 61, 67, and 75; "—" indicates data not reported.

(i.e., 1,800 versus 8 million). The average difference between fish acute toxicity values ($n = 3$) is <2 times, and there is a moderate difference of 9 times between daphnid acute values ($n = 2$). In contrast, differences between green algal 96-hr EC_{50} and algal ChV measurements are >160 times and 150 times, respectively. The steep dose–response with algae seems to suggest that algae are more sensitive to low molecular weight polycationics than species with higher surface to volume ratios.

SAR with Polycationics

The SAR equations for polycationics presented here were generated from measured data shown in Table 10.7. The data were reported in terms of 100% a.i. of polymer (100% polymer solids), test dilution water with less than 2 mg/L of TOC and hardness <180 mg/L as $CaCO_3$, pH near 7, and nominal concentrations.

Charge density of polycationics is described for OPPT purposes according to percent amine-nitrogen (%a-N) because more then 99.9% of all polymers that have been submitted under Section 5 of TSCA have had their cationic group based on nitrogen. Measures of %a-N, cation equivalent weight (EQWT), and number of cations per 1,000 MW ($\#C/K$) are all equivalent expressions of cationic charge density. To convert from %a-N to EQWT or $\#C/K$, the following equations can be used:

$$1,400 \div \%\text{a-N} = \text{N-EQWT or cation EQWT}$$

and:

$$\%\text{a-N} \times 0.714286 = \#C/K$$

To convert cation EQWT to $\#C/K$:

$$1,000 \div \text{EQWT} = \#C/K$$

From information provided in Table 10.7, SAR equations were developed here as follows, assuming all EC (e.g., LC_{50} and ChV) values are in units of mg/L:

A. Fish acute toxicity: 96-hr LC_{50}
   1. Carbon-based backbone
      (a) %a-N $\leqslant 3.5$:

$\log$ [fish 96-hr LC_{50}] = 1.209 $-$ 0.462 %a-N, where n = 19 and R^2 = 0.66

(b) %a-N > 3.5%:

fish 96-hr LC_{50} = 0.280 mg/L, where n = 34 (geometric mean LC_{50}; no correlation with charge density)

2. Silicone-based backbone
   (a) %a-N $\leqslant$ 3.5:

$\log$ [fish 96-hr LC_{50}] = 2.203 $-$ 0.963 %a-N, where n = 4 and R^2 = 0.73

(b) %a-N > 3.5%:

fish 96-hr LC_{50} = 1.17 mg/L, where n = 1

3. Natural-based backbone
   (a) %a-N $\leqslant$ 3.5: Natural-based polymers will have either similar or less toxicity when compared to carbon-based polycationics. Polymers based on tannin (polymers 77 and 78) have similar toxicity, polymers based on chitin or glucosamine (polymers 79) are 2,000 times less toxic, and polymers based on starch (polymers 71 to 76) are 10 times more toxic to 80 times less toxic. SAR analysis for polycations with natural-based polymer backbones will require the nearest analog method to predict fish acute toxicity.
   (b) %a-N > 3.5%: The only datum available (polymer 79) indicates less toxicity than when predicted with SARs for carbon-based polycationics.
B. Daphnid acute toxicity: 48-hr LC_{50}
   1. Carbon-based backbone
      (a) %a-N $\leqslant$ 3.5:

$\log$ [daphnid 48-hr LC_{50}] = 2.839 $-$ 1.194 %a-N, where n = 7 and R^2 = 0.90

(b) %a-N > 3.5%:

daphnid 48-hr $LC_{50} = 0.100$ mg/L, where $n = 13$ (geometric mean LC_{50}; no correlation with charge density)

2. Silicone-based backbone: Data for silicone-based polymers (see Table 10.7) indicate that the acute toxicity toward daphnids will be similar or less compared to carbon-based polycationics. SAR analysis will require the nearest analog method to predict daphnid acute toxicity, using the most toxic nearest analog.

3. Natural-based backbone
   (a) %a-N $\leqslant$ 4.3:

$$\log [\text{daphnid 48-hr } LC_{50}] = 2.77 - 0.412 \text{ %a-N, where } n = 6 \text{ and } R^2 = 0.82$$

   (b) %a-N > 4.3%:

$$\text{daphnid 48-hr } LC_{50} \leqslant 11 \text{ mg/L, where } n = 1$$

Cationic polymers with natural-based backbones are generally less toxic than predicted relative to carbon-based polycationics with the same charge density.

C. Green algal toxicity: 96-hr EC_{50}
   1. Carbon-based backbone
      (a) %a-N $\leqslant$ 3.5:

$$\log [\text{green algal 96-hr } EC_{50}] = 1.569 - 0.97 \text{ %a-N, where } n = 5 \text{ and } R^2 = 0.54$$

      (b) %a-N > 3.5%:

$$\text{green algal 96-hr } EC_{50} = 0.040 \text{ mg/L, where } n = 12 \text{ (geometric mean } EC_{50}; \text{ no correlation with charge density)}$$

   2. Silicone-based backbone: Data for silicone-based polymers (Table 10.7) indicate that toxicity toward green algae will be either similar to or less than that for carbon-based polycationics. SAR analysis will require the nearest analog method to predict the algal 96-hr EC_{50} toxicity value, using the most toxic nearest analog.

3. Natural-based backbone: Cationic polymers with natural-based backbones are less toxic than predicted relative to carbon-based polycationics with the same charge density. SAR analysis will require the nearest analog method, using the most toxic nearest analog.

D. Fish chronic toxicity: Only one polymer (polymer 82) has been tested for fish chronic toxicity using a fish early life-stage toxicity test. The chronic value (ChV) was 0.018 mg/L, which resulted in an acute to chronic ratio (ACR) of 18.

E. Daphnid chronic toxicity: Only one polymer (polymer 82) has been tested for daphnid chronic toxicity using the daphnid 21-day reproductive inhibition toxicity test. The ChV was 0.022 mg/L, resulting in an acute to chronic ratio (ACR) of 14.

F. Green algal chronic toxicity: 96-hr ChV
   1. Carbon-based backbone
      (a) %a-N $\leqslant$ 3.5:

      $$\log \, [\text{green algal ChV}] = 1.057 - 1 \, \%\text{a-N, where } n = 5 \text{ and } R^2 = 0.53$$

      (b) %a-N > 3.5%:

      green algal 96-hr ChV $= 0.020$ mg/L, where $n = 11$ (geometric mean ChV; no correlation with charge density)

   2. Silicone-based backbone: The single datum for silicone-based polymers (Table 10.7, polymer 62) indicated that the algal ChV was similar to algal ChVs for carbon-based polycationics. SARs for carbon-based polycationics may be used to predict the ChV for silicone-based polymers until more test data are obtained.

   3. Natural-based backbone: Cationic polymers with natural-based backbones are less toxic than predicted relative to carbon-based polycationics with the same charge density. SAR analysis will require the nearest analog method, using the most toxic nearest analog.

Amphoteric Polymers

Amphoteric polymers contain both cationic and anionic moieties (e.g., carboxylic acids) in the same polymer. The aquatic toxicity of polyamphoterics is determined primarily by the cation to anion ratio (CAR) and the cationic charge density. As cationic charge density increases, the toxicity to aquatic organisms increases. Further, when charge density is constant, toxicity increases with CAR.

When measured data are not available, available data suggest that the aquatic toxicity of a polyamphoteric may be predicted by SAR in four steps:

1. The %a-N and the CAR are calculated from the polymer's chemical structure.
2. A first estimate of the aquatic toxicity of the polyamphoteric is predicted from charge density by assuming that the subject polymer is only a carbon-based polycationic with %a-N identical to that of the polyamphoteric.
3. A toxicity reduction factor (TRF) is calculated for the polyamphoteric for each effect endpoint (e.g., daphnid acute toxicity; fish acute toxicity, etc.). The TRF is based on the polyamphoteric's CAR (see explanation below).
4. The predicted toxicity value from step 2 is multiplied by the TRF to generate the final toxicity value for the polyamphoteric polymer.

TRF equations according to polyamphoteric CAR values were developed here by SAR. Measured toxicity values for polyamphoterics are found in Table 10.8. These measured values were divided by predicted toxicity values via step 2 (see above) to provide TRF values for each Table 10.8 polyamphoteric. To develop a SAR, the same TRF values were plotted versus Table 10.8 CAR values. According to available data, SARs by linear

Table 10.8. Aquatic toxicity data for polyamphoteric polymers

Polymer no.	A-N %	$MW_n/<1,000/<500$	Cation to anion ratio	F96	D48	GA96	GAChV
				(mg/L)			
80	0.69	100K/1/1	1.3	53.0	—	—	—
81	4.7	50K/1/0	0.91	6.6	19.8	1.35	0.49
82	6.6	1,000K/0/0	5.0	0.28	0.30	0.60	—
82	6.6	1,000K/0/0	5.0	0.38	—	1.24	—
82	6.6	1,000K/0/0	5.0	0.60	—	—	—

[a]Polymer no. = polymer identification number; A-N = cationic charge density as % amine nitrogen; $MW_n/<1,000/<500$ = average-number molecular weight of polymer in thousands (K) with percent less than 1,000 and percent less than 500; Cation-to-anion ratio (CAR); and effective concentration (EC) for F96 = fish 96-hr LC_{50}, D48 = daphnid 48-hr LC_{50}, GA96 = green algal 96-hr EC_{50}, and GAChV = green algal Chronic Value.

[b]All ECs are based on 100% active ingredients or 100% polymer solids, nominal concentrations, and TOC of dilution water was less than 2 mg/L.

[c]Replicate data are given for polymer 82; "—" indicates data not reported.

regression between TRF and CAR were:

A. Fish acute toxicity: 96-hr LC_{50}:

$$\log [TRF] = 1.411 - 0.257 \, CAR, \text{ where } n = 3 \text{ and } R^2 = 0.86$$

B. Daphnid acute toxicity: 48-hr LC_{50}:

$$\log [TRF] = 2.705 - 0.445 \, CAR, \text{ where } n = 2$$

C. Green algal toxicity: 96-hr EC_{50}:

$$\log [TRF] = 1.544 - 0.049 \, CAR, \text{ where } n = 2$$

D. Green algal chronic toxicity: 96-hr ChV:

$$\log [TRF] = 1.444 - 0.049 \, CAR, \text{ where } n = 2$$

Cationic and Amphoteric Polymers: Mitigation of Toxicity

From 1970 to 1984, many polycationics submitted under Section 5 of TSCA as new chemicals were assessed as having the potential to present an unreasonable risk to the aquatic environment. As a result, many polycationics were regulated by restricting uses and/or release sites, requiring on-site treatment in addition to sewage treatment, and requiring toxicity and fate testing. The impact on the polymer manufacturing industry became so great that in 1984 the Synthetic Organic Chemicals Manufacturers Association (SOCMA) began discussions with the Environmental Effects Branch (EEB) of OPPT about the environment risk and fate of polycationics. SOCMA indicated (and it was generally known and accepted by OPPT) that polycationics were designed to react with DOC and/or suspended solids in water to form neutral insoluble complexes (i.e., floc). The floc was expected to settle out from the water column. Further, there was general information in the early literature (1960–1970s) that the toxicity of polycationics toward fish was reduced in the presence of suspended solids in water. It was proposed that polycationics should not be toxic to aquatic organisms as long as there was sufficient DOC in natural waters and/or wastewater to satisfy the ionic exchange capacity of the polymer.

It quickly became apparent to SOCMA and OPPT that fate studies would be of little use in supplying adequate test data to refine the PEC in the environment because (1) analytical methods for polycationics had higher detection limits than estimated CCs, and (2) even the best available analytical method could not distinguish between dissolved and DOC-reacted polymer. OPPT wanted adequate rigorous testing data to prove that polycationics were of low toxicity in the presence of DOC. In addition, OPPT wanted to know if polycationics in sediment were bioavailable to organisms that fed on sediment.

In 1986, EEB recommended that OPPT consider development of a TSCA Section 4 test rule for existing polycationics because discussions with SOCMA had resulted in no significant progress on testing, and it was known that there were many unpublished toxicity studies for polycationics on the Inventory. A test rule was intended to automatically generate Section 8(a) and 8(d) rules to obtain industry data on the manufacturing of polycationics and those unpublished studies. Further, OPPT began to design a standard test guideline to measure toxicity of cationic chemicals in the presence of DOC. Fish were selected by OPPT as the test species because it was expected that fish were to be less susceptible than algae and daphnids to the physical toxicity of floc caused by the reaction of poly-cationics with DOC.

The two most important factors during test development were (1) what DOC to use in the test and (2) the mean concentration and distribution of DOC in U.S. surface freshwaters. Based on research done by Cary et al. (1987), humic acid was selected by OPPT as the representative DOC for natural waters. Cary et al. measured the acute toxicity of four polycationics to freshwater fish and aquatic invertebrates in the presence of four sus-pended solids and five types of purified DOC compounds (i.e., lignin, tannic acid, fulvic acid, lignosite, and humic acid). OPPT analysis of data from Cary et al. indicated that mitigation factors (MF) for humic acid were closest to the mean MF obtained from all DOCs tested. The MF values calculated by OPPT were ratios of measured acute toxicity LC_{50} values (per species tested) in water with added DOC or suspended solids to measured acute toxicity LC_{50} values in clean dilution water where TOC $< 2\,mg/L$. The order of difference from the mean MF was: lignin $>$ tannic acid $>$ fulvic acid $>$ lignosite $>$ humic acid.

Humic acid is readily available for use in standard laboratory tests; however, concentrations of humic acid in natural waters are rarely meas-ured. TOC may be used as a practical surrogate to determine appropriate test levels of DOC as humic acid. Lynch (1987) analyzed the U.S. EPA Office of Water's STORET database for measured amounts of TOC in U.S. waters and found 67,994 measurements of TOC taken from 1977 through 1987 from 19 of 23 major river basins in the United States. The geometric mean of these data was 6.8 mg/L TOC. The benchmark of 10 mg/L TOC with humic acid approximates 6.8 mg/L TOC and is considered to be a median estimate of TOC. In the STORET database, TOC levels were log-normally distributed and skewed toward larger concentrations of TOC than 6.8 mg/L TOC.

By November 1988, negotiations between SOCMA and OPPT resulted in the OPPT test guideline Fish Acute Toxicity Mitigated by Humic Acid (OPPTS 850.1085). The guideline was distributed in December 1988. In October 1989, seven acute toxicity and toxicity mitigation tests using fish had been submitted to OPPT and validated by EEB. As a result, EEB recommended to OPPT that the proposed Section 4 test rule and Section 8(a) and Section 8(d) rules for polycationics be withdrawn. There is no current activity on these polymers in the OPPT Existing Chemical Program.

As shown in Table 10.9, the toxicity of polycationics is mitigated in the presence of DOC. Fish acute toxicity data for 16 polycationics have been evaluated and validated by EEB. After these polymers were submitted under Section 5 as new chemicals, they were dropped from further review after the MF was integrated into the risk assessment. In Table 10.9, MF was defined as the predicted fish 96-hr LC_{50} value in water with 10 mg/L TOC added to clean (i.e., TOC < 2 mg/L) dilution water divided by the measured fish 96-hr LC_{50} value in clean dilution water. The 96-hr LC_{50} value in water with 10 mg/L TOC added was predicted by linear regression of each measured LC_{50} with TOC.

MF values for polycationics that (1) are random reactions of monomers and (2) have minimal oligomer content (i.e., $<25\% < 1,000$ and $< 10\% < 500$) are correlated with charge density (i.e., %a-N). From currently available information, SARs developed here are:

A. For charge densities $\geqslant 3.5$ %a-N:

$$MF = 110, \text{ where n} = 7$$

B. For charge densities between 3.5 and 0.7 %a-N:

$$\log [MF] = 0.858 + 0.265 \text{ %a-N, where } n = 4 \text{ and } R^2 = 0.61$$

C. For charge densities <0.7 %a-N: MFs have not been measured, but are expected to be <7.

Polycationics that have significant oligomer content (i.e., $>25\% <$ 1,000 and $>10\% < 500$) can have MFs that are significantly lower than

Table 10.9. Mitigation of acute toxicity in fish of polycationics and polyamphoterics by humic acid[a,b]

Polymer no.	A-N (%)	MW_n / < 1,000 / < 500	TOC (mg/L)	Fish 96-hr LC_{50} (mg/L)	Mitigation[c] factor
27	0.69	6.2K/0/0	1.9	4.6	—
27	0.69	6.2K/0/0	4.1	82.0	—
27	0.69	6.2K/0/0	7.5	350	130
29	0.70	8,000K/0/0	0.45	3.3	—
29	0.70	8,000K/0/0	3.2	8.6	—
29	0.70	8,000K/0/0	5.9	14.7	7.0
33	1.4	2.4K/13/2	1.0	17.0	—
33	1.4	2.4K/13/2	2.5	19.0	—
33	1.4	2.4K/13/2	5.9	190	—
33	1.4	2.4K/13/2	9.8	190	24.0
69	2.6	1.7K/17/5	1.5	0.65	—
69	2.6	1.7K/17/5	10.0	30.0	—
69	2.6	1.7K/17/5	20.0	84.0	65.0
39	3.3	5,000K/0/0	1.9	1.5	—
39	3.3	5,000K/0/0	4.8	12.0	—
39	3.3	5,000K/0/0	14.0	58.0	35.0
78	3.4	1.7K/30/25	1.5	1.5	—
78	3.4	1.7K/30/25	5.0	30.0	—
78	3.4	1.7K/30/25	12.0	110	71.0
42	3.6	250K/0/0	0.0	0.73	—
42	3.6	250K/0/0	1.5	7.5	—
42	3.6	250K/0/0	3.9	13.0	48.0
43	4.5	1,000K/0/0	0.0	0.31	—
43	4.5	1,000K/0/0	0.250	5.9	—
43	4.5	1,000K/0/0	1.2	17.0	—
43	4.5	1,000K/0/0	5.1	46.0	250
44	4.6	1,000K/0/0	0.0	0.90	—
44	4.6	1,000K/0/0	0.250	6.0	—
44	4.6	1,000K/0/0	1.2	20.0	—
44	4.6	1,000K/0/0	5.1	40.0	76.0
47	6.4	1K/38/12	0.0	0.72	—
47	6.4	1K/38/12	3.3	5.2	—
47	6.4	1K/38/12	5.5	8.9	21.0
82	6.6	1,000K/0/0	0.0	0.28	—
82	6.6	1,000K/0/0	0.0	0.38	—
82	6.6	1,000K/0/0	0.0	0.60	—
82	6.6	1,000K/0/0	0.250	3.6	—
82	6.6	1,000K/0/0	0.250	3.6	—
82	6.6	1,000K/0/0	0.570	6.1	—
82	6.6	1,000K/0/0	1.2	12.2	—
82	6.6	1,000K/0/0	2.5	28.0	—
82	6.6	1,000K/0/0	5.1	46.0	290

Table 10.9.   Continued

Polymer no.	A-N (%)	$MW_n/<1,000/<500$	TOC (mg/L)	Fish 96-hr LC_{50} (mg/L)	Mitigation[c] factor
48	7.8	1K/32/23	1.0	1.0	—
48	7.8	1K/32/23	3.7	10.0	—
48	7.8	1K/32/23	7.1	19.0	26.0
83	8.4	100K/0/0	5.2	72.0	—
83	8.4	100K/0/0	7.2	440	—
83	8.4	100K/0/0	13.0	1,456	27.0
54	11.0	1.5K/44/26	1.0	0.20	—
54	11.0	1.5K/44/26	1.7	>2.0	—
54	11.0	1.5K/44/26	3.4	>2.0	—
54	11.0	1.5K/44/26	6.9	>20.0	—
54	11.0	1.5K/44/26	14.0	>20.0	>180
56	14.0	2K/20/3	0.0	0.10	—
56	14.0	2K/20/3	0.0	0.16	—
56	14.0	2K/20/3	1.5	3.0	—
56	14.0	2K/20/3	3.1	5.36	—
56	14.0	2K/20/3	6.3	10.2	140
58	14.0	2K/25/5	0.0	0.072	—
58	14.0	2K/25/5	0.0	0.084	—
58	14.0	2K/25/5	0.0	0.16	—
58	14.0	2K/25/5	1.5	2.2	—
58	14.0	2K/25/5	3.1	5.4	—
58	14.0	2K/25/5	6.3	9.6	170

[a]Polymer no. = Polymer identification number; A-N = cationic charge density in % amine nitrogen; $MW_n/<1,000/<500$ = average-number molecular weight of polymer in thousands (K) with percent less than 1,000 and percent less than 500.

[b]All ECs are based on 100% active ingredients or 100% polymer solids, nominal concentrations, and pH approximately 7.

[c]Mitigation factor = the predicted fish 96-hr LC_{50} value in water with 10 mg/L TOC divided by the fish 96-hr LC_{50} value in clean (i.e., TOC < 2 mg/L) dilution water. See text for details.

predicted for polymers with minimal oligomer content. For example, polymers 47 and 48 have predicted MFs of 110; however, their measured MFs were 21 and 26, respectively. In addition, MFs for blocked polymers cannot, at this time, be accurately predicted by using these SARs. Polymer 27 is a blocked polymer with the cationic group attached to the polymer ends. Except for polymer 27, polymers in Table 10.9 are randomly reacted polymers. The measured MF was 130, whereas the predicted MF was 11.

    The aquatic toxicity of polycationics is also mitigated when the polymers are mixed with sediment. Available toxicity test data with natural sediment contaminated with polycationics and with benthic species that ingest sediment has shown that polycationics with charge densities of $\geqslant 4.2$ %a-N (or $\geqslant 3$ cations/1,000 MW or an N-equivalent weight $\leqslant 333$) are not bioavailable to cause toxicity. The geometric mean of the 48-hr LOEC and the 48-hr NOEC was $> 100 \, mg/kg$ of dry weight sediment (Rogers and Witt, 1989), suggesting low concern for toxicity when they are transported to sediments.

A word of caution is needed here. No polycationic with significant oligomer content (i.e., $> 25\% < 1,000$ and $> 10\% < 500$) has been tested. Oligomers are expected to be more bioavailable, and lower MFs are observed for polymers with significant oligomer content (Table 10.9). Another exception may occur when a polycationic is formulated with acid in excess of the amount needed to neutralize the polymer, such that the product's pH is approximately 2. At low pH, polycationics can fail to flocculate DOC, and DOC apparently does not mitigate acute toxicity to fish (Nabholz et al., 1993a).

OPPT often has sufficient data to predict the mitigation of many polycationics. However, data gaps may exist. For example, in the absence of suitable mitigation test data, a useful testing scheme is: (1) fish acute toxicity with clean dilution water (i.e., TOC $< 2 \, mg/L$), (2) daphnid acute toxicity with clean dilution water, (3) algal toxicity test with standard growth medium, (4) fish acute toxicity with 10 mg of humic acid/L added to clean dilution water, and (5) fish acute toxicity with 20 mg humic acid/L added to dilution water. All tests can be done under static conditions with nominal concentrations of polymer.

When the test reults are validated by EEB, the three fish 96-hr LC_{50} values are plotted (i.e., regressed) against TOC concentration. The fish 96-hr LC_{50} from 10 mg/L TOC can be predicted from the regression equation. To obtain the MF, this value is divided by the fish LC_{50} value measured in clean water. The MF is used to adjust upward all clean water EC values in the polymer's toxicity profile. A new CC then is determined for the polymer by assuming the presence of 10 mg/L TOC. If there is still a potential risk to aquatic organisms in the water column, chronic toxicity tests in clean dilution water with fish and invertebrates under flow-through or static-renewal conditions (with renewals every 24 hr) and nominal concentrations are recommended.

CONCLUSIONS

In the United States, the environmental safety of chemical substances with nonfood, nondrug, and nonpesticidal uses is evaluated by U.S. EPA under the authority granted by TSCA. For polymers, the scope of TSCA's applicability is very broad and includes such use categories as water and wastewater treatment, coatings, household and industrial cleaning products, and manufactured goods. TSCA distinguishes between existing chemical substances and new substances not on the TSCA Inventory. To date most of the environmental assessments for polymers have focused on new polymers, for which prospective manufacturers or importers must file a PMN under Section 5 of TSCA. The PMN process essentially involves weighing potential risks and benefits for each new substance.

Polymers that meet certain requirements are exempt from reporting under Section 5, but this exemption does not include most polymers with uses such as water treatment applications. Nonexempt polymers are subject to the normal PMN review process, which includes chemistry review, toxicity evaluation, exposure assessment, and risk assessment/risk management phases. A unique feature of this process is its high level of reliance on SAR analysis to predict missing data. Spurred by the need to make informed judgments often with few or no experimental data, SAR-based estimation methods have been developed for most physical-chemical properties, some transformation processes (e.g., biodegradation), and many toxic effects relevant to aquatic exposures. Unfortunately, these methods, although on balance successful for discrete (class 1) substances, are of little benefit for predicting the fate of polymers. Measured fate data continue to play a critical role in environmental fate assessment of polymers.

Most polymers in commerce are essentially nonbiodegradable. Environmental fate assessment, therefore, is reduced to an evaluation of their potential for sorption or precipitation under various conditions. Default assumptions help to make judgments about the importance of these processes, but such assumptions are generally based on few measured data. It is in industry's best interest to furnish U.S. EPA with relevant measured fate data. Information presented herein in Chapters 6 through 9 is particularly useful for further understanding of the environmental fate of polymers.

SAR methods for predicting aquatic toxicity are available for major classes of polymers. Ecotoxicity assessment of polymers can be based on polymer charge: anionic, nonionic, cationic, and amphoteric:

- *Anionic:* Polyanionics are divided into two classes: poly(aromatic acids) and poly(aliphatic acids). Many poly(aromatic acids) based on sulfonate or carboxylate groups are moderately toxic to aquatic organisms by an unknown mechanism of action. The nearest analog SAR method is presently used by OPPT to predict the toxicity of these polymers by identifying the dominant monomer(s) of the polymer. Poly(aliphatic acids) show moderate toxicity only toward algae. These polymers may indirectly affect algal growth by chelating nutrient elements needed for growth, as the potency of such polymers is directly related to the ability to chelate divalent metals. Available data suggest that the distance between acids on the polymers affects the degree of algal toxicity, and algal toxicity can be predicted by determining the average distance between acids. Algal toxicity of poly(aliphatic acids) can be mitigated by changing their salt to a divalent ion (e.g., Ca^{2+}), or by testing with moderately hard water (e.g., 120 mg/L as $CaCO_3$). Ecological risk from these polymers will most likely be observed only in rare instances within soft natural waters with hardness of < 30 mg/L as $CaCO_3$.

- *Nonionic:* Nonionic polymers are generally of low concern for ecotoxicity; however, some nonionic polymers for use as dispersants on surfactants can be toxic to aquatic organisms. Likewise, polymers with significant amounts of oligomers whose $MW < 1,000$ may be of concern because of the possible toxicity of the oligomers; nonionic oligomers are assessed as monomers.

- *Cationic:* Polycationics are grouped by the type of polymer backbone: carbon-based, silicone-based, and backbones using natural polymers such as starch. They can be highly toxic to aquatic organisms and should be used with caution. However, mitigation in the presence of DOC is possible. The aquatic toxicity of polycationics is primarily related to their cationic charge density, and, if they are soluble or dispersible as emulsions, their aquatic toxicity can be reasonably predicted by SAR. With possible exceptions, the aquatic toxicity of silicone-based and natural-based polycationics may be less than the aquatic toxicity of carbon-based polycationics.

- *Amphoteric:* Polyamphoteric polymers contain both cationic and anionic moieties in the same polymer. The aquatic toxicity of these polymers is determined by cation-to-anion ratio as well as cationic charge density and may be mitigated by DOC. If cationic and amphoteric polymers are used according to current standard application practices, they are of low ecological risk (see also Chapter 7).

Environmentally safe manufacture and use of polymers is best ensured if U.S. EPA and industry cooperate to the extent and from the earliest phase of product development practicable for the PMN submitter. Use of the New Chemical Program's pre-notice communication process could be greater than is the case now, and the effect can only be beneficial for all stakeholders. The most important information needed by U.S. EPA assessors is polymer chemical structure. Specific data about polymer structure is crucial to the entire assessment process. Data needs include, but are not limited to, monomer identity, polymer composition (e.g., mole ratios of monomers), reaction sequence of monomers (random or blocked), MW_n, molecular weight distribution, and oligomer content (i.e., $\% < 1,000$ and $\% < 500$).

If testing is considered necessary, either by voluntary agreement or Consent Order, test protocols should be reviewed by U.S. EPA to resolve uncertainties before ecological testing starts. What is being tested should be understood and agreed to by the stakeholders. Through a combination of testing tailored to specific substances subject to PMN review and more fundamental research on the mechanisms of sorption, precipitation, toxicity, and mitigation, the database needed for truly informed ecological risk assessment with polymers can be realized.

DISCLAIMER

This chapter has been reviewed by U.S. EPA and approved for publication. Approval does not signify that the contents necessarily reflect the views and policies of the Agency, nor does mention of tradenames or commercial products constitute endorsement or recommendation for use.

REFERENCES

Alexander, M. 1973. Nonbiodegradable and other recalcitrant molecules. *Biotechnology and Bioengineering* 15:611–47.
ASTM. 1994. Standard practice for coagulation–flocculation jar test of water, D 2035-80, in *1994 Annual Book of ASTM Standards,* Section 11, Vol. 11.02. Philadelphia, PA: American Society for Testing and Materials.
Bailey, W. J. and G. Gapud. 1986. Synthesis of new biodegradable polymers, in *Biodeterioration 6. Proceedings of the Sixth International Biodeterioration Symposium,* ed. S. Barry and D. R. Houghton. Slough, U.K.: CAB International.
Barnes, D. G. and M. L. Dourson. 1988. Reference dose (RfD): Description and use in health risk assessment. *Regulatory Toxicology and Pharmacology* 8:471–86.
Biesinger, K. E. and G. N. Stokes. 1986. Effects of synthetic polyelectrolytes on selected aquatic organisms. *Journal of the Water Pollution Control Federation* 58:207–13.

Cary, G. A., J. A. McMahon, and W. J. Kuc. 1987. The effect of suspended solids and naturally occurring dissolved organics in reducing the acute toxicities of cationic polyelectrolytes to aquatic organisms. *Environmental Toxicology and Chemistry* 6:469–74.

Clark, B., J. G. Henry, and D. Mackay. 1995. Fugacity analysis and model of organic chemical fate in a sewage treatment plant. *Environmental Science and Technology* 29:1488–94.

Clements, R. G. 1988. *Estimating Toxicity of Industrial Chemicals to Aquatic Organisms Using Structure Activity Relationships.* U.S. EPA Report No. 560/6-88-001. Washington, DC: Office of Pollution Prevention and Toxics, U.S. Environmental Protection Agency.

Clements, R. G. and J. V. Nabholz. 1994. *ECOSAR: A Computer Program for Estimating the Ecotoxicity of Industrial Chemicals Based on Structure Activity Relationships.* U.S. EPA Report No. 748-R-93-002. Washington, DC: Office of Pollution Prevention and Toxics, U.S. Environmental Protection Agency.

Clements, R. G., J. V. Nabholz, D. E. Johnson, and M. G. Zeeman. 1993. The use of quantitative structure–activity relationships (QSARs) as screening tools in environmental assessment, in *Environmental Toxicology and Risk Assessment*, 2nd Volume, ASTM STP 1216, ed. J. W. Gorsuch, F. J. Dwyer, C. G. Ingersoll, and T. W. LaPoint. Philadelphia, PA: American Society for Testing and Materials.

Halverson, F. and H. P. Panzer. 1980. Flocculating agents, in *Kirk-Othmer Encyclopedia of Chemical Technology*, Vol. 10, 3rd Ed. New York: John Wiley & Sons.

HCN. 1989. *Assessing the Risk of Toxic Chemicals for Ecosystems.* The Hague, The Netherlands: Health Council of the Netherlands.

Leo, A. and D. Weininger. 1985. CLOGP Version 3.3: Estimation of the *n*-octanol/water partition coefficient for organics in the TSCA Industrial Inventory. Claremont, CA: Pomona College.

Lynch, D. G. 1987. Summary of STORET data on dissolved organic carbon (DOC) levels in surface waters. Washington, DC: Exposure, Economics, and Technology Division, Office of Pollution Prevention and Toxics, U.S. Environmental Protection Agency.

Lynch, D. G., N. F. Tirado, R. S. Boethling, G. R. Huse, and G. C. Thom. 1991. Performance of on-line chemical property estimation methods with TSCA Premanufacture Notice chemicals. *Ecotoxicology and Environmental Safety* 22:240–49.

Meylan, W. M. and P. H. Howard. 1991. Bond contribution method for estimating Henry's law constants. *Environmental Toxicology and Chemistry* 10:1283–93.

Meylan, W. M. and P. H. Howard. 1995. Atom/fragment contribution method for estimating octanol–water partition coefficients. *Journal of Pharmacological Science* 84:83–92.

Meylan, W. M., P. H. Howard, and R. S. Boethling. 1992. Molecular topology/fragment contribution method for predicting soil sorption coefficients. *Environmental Science and Technology* 26:1560–67.

Meylan, W. M., P. H. Howard, and R. S. Boethling. 1996. Improved method for estimating water solubility from octanol/water partition coefficient. *Environmental Toxicology and Chemistry* 15:100–106.

Moss, K., D. Locke, and C. Auer. 1996. EPA's new chemicals program. *Chemical Health and Safety* (Jan./Feb.):29–33.

Nabholz, J. V. 1991. Environmental hazard and risk assessment under the United States Toxic Substances Control Act (TSCA). *Science of the Total Environment* 109/110:649–65.

Nabholz, J. V., P. Miller, and M. Zeeman. 1993a. Environmental risk assessment of new chemicals under the Toxic Substances Control Act (TSCA) Section 5, in *Environmental Toxicology and Risk Assessment*, ASTM STP 1179, ed. W. G. Landis, J. S. Hughes, and M. A. Lewis. Philadelphia, PA: American Society for Testing and Materials.

Nabholz, J. V., R. G. Clements, M. G. Zeeman, K. C. Osborn, and R. Wedge. 1993b. Validation of structure activity relationships used by the U.S. EPA's Office of Pollution Prevention and Toxics for the environmental hazard assessment of industrial chemicals, in *Environmental Toxicology and Risk Assessment*, 2nd Volume, ASTM STP 1216, ed. J. W. Goruch, F. J. Dwyer, C. G. Ingersoll, and T. W. LaPoint. Philadelphia, PA: American Society for Testing and Materials.

Nalco Chemical Company. 1988. *The NALCO Water Handbook.* New York: McGraw-Hill Book Company.

NRC. 1994. *Science and Judgment in Risk Assessment.* National Research Council, Washington, DC: National Academy of Sciences.

Opgenorth, H. J. 1992. Polymeric materials polycarboxylates, in *The Handbook of Environmental Chemistry*, Vol. 3, Part F, Anthropogenic Compounds, ed. O. Hutzinger. Berlin: Springer-Verlag.

Park, J. K., D. Jenkins, T. M. Holsen, T. W. Warnock, and W. E. Gledhill. 1989. Fate of the detergent builder, sodium polyglyoxylate, in waste water treatment. *Journal of the Water Pollution Control Federation* 61:491–99.

Pranamuda, H., T. Tokiwa, and H. Tanaka. 1995. Microbial degradation of an aliphatic polyester with a high melting point, poly(tetramethylene succinate). *Applied and Environmental Microbiology* 61:1828–32.

Rodier, D. J. and D. A. Mauriello. 1993. The quotient method of ecological risk assessment and modeling under TSCA: A review, in *Environmental Toxicology and Risk Assessment*, ASTM STP 1179, ed. W. G. Landis, J.

S. Hughes, and M. A. Lewis. Philadelphia, PA: American Society for Testing and Materials.

Rogers, J. R., Jr. and W. T. Witt. 1989. Effects of sediments flocculated with cationic polyelectrolytes when fed upon by *Daphnia magna.* Denton, TX: Department of Biological Sciences, University of North Texas.

Smrchek, J., R. Clements, R. Morcock, and W. Rabert. 1993. Assessing ecological hazard under TSCA: Methods and evaluation of data, in *Environmental Toxicology and Risk Assessment*, ASTM STP 1179, ed. W. G. Landis, J. S. Hughes, and M. A. Lewis. Philadelphia, PA: American Society for Testing and Materials.

U.S. EPA. 1982. *Fate of Priority Pollutants in Publicly Owned Treatment Works.* U.S. EPA Report No. 440/1-82-303. Washington, DC: Office of Water, U.S. Environmental Protection Agency.

U.S. EPA. 1985. *Exposure Analysis Modeling System (EXAMS): User Manual and System Documentation.* U.S. EPA Report No. 600/3-82-023. Washington, DC: Office of Research and Development, U.S. Environmental Protection Agency.

U.S. EPA. 1988. *User's Guide to PDM3.* Washington, DC: Office of Pollution Prevention and Toxics, U.S. Environmental Protection Agency.

U.S. EPA. 1990. *Statistical Methods for Estimating Exposure Above the Reference Dose.* U.S. EPA Report No. 600/8-90-065. Washington, DC: Office of Health and Environmental Assessment, U.S. Environmental Protection Agency.

U.S. EPA. 1994a. *A Review of Ecological Assessment Case Studies from a Risk Assessment Perspective*, Vol. II, Part II, Section 1: New Chemical Case Study. U.S. EPA Report No. 630-R-94-003. Washington, DC: Office of Research and Development, U.S. Environmental Protection Agency.

U.S. EPA. 1994b. *U.S. EPA/EC Joint Project on the Evaluation of Quantitative Structure Activity Relationships.* U.S. EPA Report No. 743-R-94-001. Washington, DC: Office of Prevention, Pesticides, and Toxic Substances, U.S. Environmental Protection Agency.

U.S. EPA. 1995a. *New Chemicals Program.* U.S. EPA Report No. 743-F-95-001. Washington, DC: Office of Pollution Prevention and Toxics, U.S. Environmental Protection Agency.

U.S. EPA. 1995b. Amendment for expedited process to issue significant new use rules for selected new chemical substances. Final Rule. *Federal Register* 60:16311–16316.

U.S. EPA. 1995c. Premanufacture notification for exemption, revision of exemption for chemical substances manufactured in small quantities, low release and exposure exemption. Final Rule. *Federal Register* 60: 16336–16351.

U.S. EPA. 1995d. Premanufacture notification exemptions, revisions of

exemptions for polymers. Final Rule. *Federal Register* 60: 16316–16336.
U.S. EPA. 1996. *Polymer Exemption Guidance Manual.* Washington, DC: Office of Pollution Prevention and Toxics, TSCA Assistance Information Service, U.S. Environmental Protection Agency.
Wagner, P. M., J. V. Nabholz, and R. J. Kent. 1995. The new chemicals process at the Environmental Protection Agency (EPA): Structure–activity relationships for hazard identification and risk assessment. *Toxicology Letters* 79:67–73.
Zeeman, M. and J. Gilford. 1993. Ecological hazard evaluation and risk assessment under EPA's Toxic Substances Control Act (TSCA): An introduction, in *Environmental Toxicology and Risk Assessment*, ASTM STP 1179, ed. W. G. Landis, J. S. Hughes, and M. A. Lewis. Philadelphia, PA: American Society for Testing and Materials.

Chapter 11

Regulation of New Polymers in Canada

■ ROBERT A. MATHESON AND ANDREW J. ATKINSON
Environment Canada

INTRODUCTION

This chapter provides an overview of the system for regulation of new polymers in Canada under the Canadian Environmental Protection Act (CEPA). It is not an in-depth description of the regulations or a step-by-step guidance document. Requirements specific to "Transitional Substances" (substances imported or manufactured in quantities greater than 20 kg during a calendar year between January 1, 1987 and June 30, 1994) have not been addressed here. Individuals required to notify a polymer in Canada are directed to the Act and Regulations as well as to the publication *Guidelines for the Notification and Testing of New Substances, Chemicals and Polymers* (SSC, 193).

As part of an overall "cradle to grave" management approach to toxic substances, CEPA includes provisions specific to "Substances New to Canada." The CEPA approach to control of new substances is preventive in nature. A pre-manufacture/pre-import assessment process is utilized to ensure that potentially problematic chemicals destined for the Canadian marketplace are identified and appropriate control actions implemented.

New substances are jointly assessed by the Canadian federal Departments of Environment and Health to determine the potential adverse effects of the substance on the environment and human health. This assessment results in:

(a) a determination that the substance is not suspected of being toxic; or

(b) a determination there is a suspicion that the substance is toxic, which may require: (1) controls or prohibition, or (2) prohibition pending submission and assessment of additional information determined to be required by the Departments.

As defined within the statute, a substance is considered "toxic" if it is entering or may enter the environment in a quantity or concentration or under conditions that:

(a) have or may have an immediate or long-term harmful effect on the environment,
(b) constitute or may constitute a danger to the environment on which human life depends, or
(c) constitute or may constitute a danger in Canada to human life or health.

POLYMERS SUBJECT TO NOTIFICATION

Polymer Definition

New Substances Notification Regulations developed under the authority of CEPA were published in the *Canada Gazette*, Part II on April 6, 1994 and were effective July 1 of the same year. Under the regulatory framework, polymers are dealt with as a particular generic class of substance and are assessed under Part II of the Regulations (New Substances That Are Polymers).

For the purpose of substances identification, the definition of a polymer developed by the Organisation for Economic Co-operation and Development (OECD) is used. Substances are considered polymers for the purpose of notification if they have all four of the following characteristics:

(a) Molecules are characterized by the sequence of one or more types of monomer units.
(b) More than 50% of the molecules (by weight) contain at least three monomer units covalently bound to at least one other monomer or reactant.
(c) Less than 50% of the molecules (by weight) are of the same molecular weight.
(d) Molecules are distributed over a range of molecular weights, with differences in the molecular weights primarily attributable to differences in the number of monomer units.

Identifying New Polymers

Under CEPA, a Domestic Substances List (DSL) has been compiled to serve a regulatory role similar to that of the Toxic Substances Control Act (TSCA) Inventory in the United States. All polymers listed on the DSL are exempt from notification. New polymers (i.e., those requiring notification) are those not included on the list. The DSL consists of substances that either (1) were used in Canada between January 1, 1984 and December 31, 1986 or (2) have been added after an evaluation by the Departments (as a consequence of notification under the requirements of the Regulations) has not resulted in the imposition of any statutory "condition," and prescribed volumes of manufacture or import have been exceeded.

The DSL consists of an initial list published in the May 4, 1994 edition of *Canada Gazette*, Part II plus all other additions and amendments published periodically in the same publication. Polymer entries on the DSL are identified by both a Chemical Abstract Service Registry Number and a chemical name, constructed by using the Chemical Abstracts Service nomenclature convention of naming polymers in terms of their constituent monomers.

Polymers Not Requiring Notification

(a) A polymer not included on the DSL does not require notification if it is manufactured or imported for a use that is regulated under another Canadian federal statute that requires notification prior to manufacture, import, and sale and for an assessment of whether it is toxic. Examples include the Pest Control Products Act and the Food and Drugs Act.

(b) A polymer manufactured by modifying the formulation of a polymer specified on the DSL by adding reactants, none of which constitutes more than 2% by weight of the polymer, is exempt from notification. The word "modifying" refers to either the amount of additional reactant that has been incorporated into the structure of the polymer or the amount charged to the vessel.

(c) A polymer produced when a substance undergoes a chemical reaction that is incidental to the use to which the substance is put or that results from storage or from environmental factors need not be notified. Examples of incidental reaction products formed are substances formed from chemical reactions during:

    1. The blending of a formulation where there is no intention to produce new polymers, and any ensuing chemical reactions do not enhance the commercial value of the formulation.

2. Exposure to environmental factors such as air, moisture, and sunlight.
3. Storage.
4. The intended use of a substance.

(d) Polymers that are impurities, contaminants, and partially unreacted materials, the formation of which is related to the preparation of a substance, need not be notified. Impurities and contaminants are substances that are normally found in minimal concentration in the starting materials or are the result of secondary reactions that occur during the manufacturing process. These substances, and partially unreacted starting materials that are present in the final product, are the direct result of the preparation, are not necessary to the end use of the product, have not been intentionally added to the substance, and do not enhance the commercial value of the substance.

REGULATORY STRUCTURE

Influencing Factors

CEPA prohibits the import or the manufacture of any new polymer unless information, prescribed within the New Substances Notification Regulations, is provided a specified number of days prior to the exceedence of particular volume thresholds. Part II of the Regulations prescribes information requirements (Schedules VI, VII, VIII, XI, XII, and XIII) tailored to the type, use, and quantity of the polymer.

The following factors are particularly important in determining the extent of information required for evaluation of a polymer:

(a) Whether the polymer is listed on the Non-domestic Substances List (NDSL) (described below).
(b) Whether or not each monomer and reactant appears on the NDSL or the DSL.
(c) Whether the polymer meets the criteria for a low concern polymer.
(d) Whether the polymer falls within any of several special categories.
(e) Whether yearly and cumulative manufacture or import quantities will exceed regulatory specified quantities.

Figure 11.1 is a schematic representation of the decision process followed in selecting the appropriate regulatory Schedule.

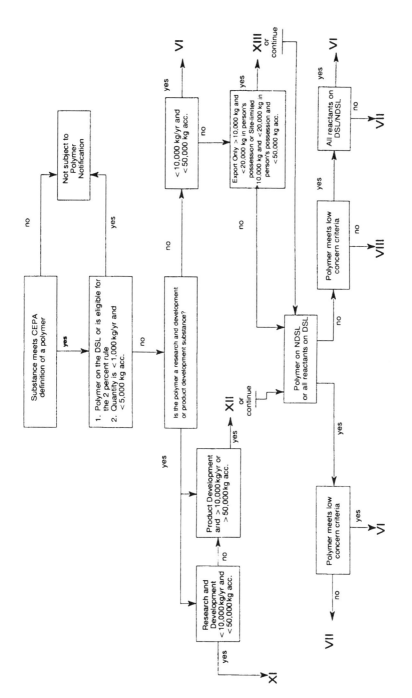

Figure 11.1. Notification requirements for new polymers.

Polymers on the Non-domestic Substances List

The Non-domestic Substances List (NDSL) specifies substances that are not on the DSL but are believed to be in international commerce. The NDSL, published in the *Canada Gazette*, Part I on January 26, 1991, consisted of the 1985 version of the U.S. TSCA Inventory minus the substances on the DSL.

Substances listed on the NDSL require less detailed notification packages for assessment than substances that are new to both the Canadian marketplace and world commerce. For substances with a use history of several years, it is assumed that (1) information can be accessed through the open literature, and (2) government agencies in other jurisdictions will most likely have conducted reviews for health and environmental impacts, and the results of such studies will be known.

Beginning in 1995, the NDSL has been revised annually to add and delete substances to reflect the composition of the TSCA Inventory five years before the year of update. For example, the 1995 NDSL revision is based on the 1990 TSCA. Substances on the TSCA Inventory that have been flagged by the U.S. EPA for the purpose of controlling manufacture, import, or use are not eligible for placement on the NDSL through this annual revision process.

Monomers and Reactants on the NDSL and DSL

In general, it is assumed that a polymer has less impact on human health and the environment than do the constituent monomers or reactants. The presence of these substances on the DSL or the NDSL may reduce the requirement for information specific to the polymer, as data on its constituents can be used in the evaluation process.

Low Concern Polymers

The criteria used to identify those polymers eligible for the U.S. TSCA Polymer Exemption Rule have been used to define a low concern polymer under the CEPA Regulations. In general, such polymers are deemed to have insignificant environmental and human health impacts, and, consequently, only minimal information is required.

Low concern polymers are identified in a two-step process. First, Schedule IX of the Regulations identifies five types of polymers that have the potential to pose a relatively higher risk to the environment and human health than other polymers:

1. A cationic polymer or a polymer that is expected to become cationic in the aquatic environmental, except for a polymer that has a combined equivalent weight for the cationic group greater than 5,000. The equivalent weight for the cationic group or the functional group equivalent weight (FGEW) is the ratio of the mass of the polymer to the number of moles of the cationic group.

2. A polymer that is designed, or can be expected, to degrade, decompose, or depolymerize.

3. A polymer that includes no more than one of the following elements: carbon, hydrogen, nitrogen, oxygen, silicon, and sulfur (i.e., a polymer that contains carbon but does not contain hydrogen, nitrogen, oxygen, silicon, and sulfur cannot be considered low concern).

4. A polymer containing:
   (a) Any elements other than carbon, hydrogen, nitrogen, oxygen, silicon, sulfur, fluorine, chlorine, bromine, or iodine;
   (b) Any monatomic counter ions other than chlorine ion, bromine ion, iodine ion, sodium ion, divalent magnesium, trivalent aluminum, potassium ion, or divalent calcium; and
   (c) Any elements other than lithium, boron, phosphorus, titanium, manganese, iron, nickel, copper, zinc, tin, or zirconium, unless any combination of these elements is present at 0.2% or more by weight.

5. A polymer that contains any reactive functional groups other than carboxylic acid groups, aliphatic hydroxyl groups, unconjugated olefinic groups, butenedioic groups, cyano groups, blocked isocyanates including ketoxime-blocked isocyanates, unconjugated nitrile groups, halogens excluding halides, and conjugated olefinic groups in naturally occurring fats, oils, and carboxylic acids, in combined amounts greater than:
   (a) One part in 5,000; or
   (b) Where the only reactive functional groups present are part of acid halides, acid anhydrides, aldehydes, hemiacetals, methylol-amides, methylol-amines, methylol-ureas, greater than C_2 alkoxysilanes, allyl ethers, conjugated olefins, cyanates, epoxides, imines, or unsubstituted positions *ortho* or *para* to phenolic hydroxyl, one part in 1,000.

The first series of functional groups stated in this item are those not considered to be reactive; polymers containing only these reactive groups are candidates for low concern status. Part (a) consequently applies only to reactive groups *not* mentioned in the first series of functional groups noted under this item. Reactive groups in combined amounts less than or equal to an FGEW of 5,000 do not meet the low concern criteria. Part (b) lists a second series of reactive functional groups. Polymers that contain only these groups in combined amounts less than or equal to an FGEW of 1,000 also are not eligible for low concern status.

The second step is specifically to identify low concern criteria. These include polymers of high number-average molecular weight that have a limited percentage of low molecular weight components or are specific polyesters.

Low concern polymers are polymers that:

(a) Are not described in items 1, 2, 3, or 4 above and have a number-average molecular weight greater than 10,000 less than 2% of components have a molecular weight less than 500, and less than 5% of components have a molecular weight less than 1,000 (polymers described in item 5 are eligible for consideration);

(b) Are not described in any of the five items and have a number-average molecular weight greater than 1,000, less than 10% of components have a molecular weight less than 500, and less than 25% of components have a molecular weight less than 1,000; or

(c) Are polyesters manufactured solely from reactants on the DSL and listed in Schedule X of the Regulations unless the polyester is manufactured from a combination of reactants that includes both 1-butanol and (E) 2-butenedioic acid.

Special Categories of Polymers

Special categories of polymers have been defined to identify situations where there is a reduced likelihood of environmental exposure and, therefore, a reduced requirement for information.

Research and Development Polymers

Research and development polymers are subject of systematic investigation or search (by means of experimentation or analysis, or both) that has as its primary objective the creation or the improvement of a product. This category includes polymers being manufactured on a fee basis for domestic or foreign customers that are conducting research. These polymers are subject to reduced information requirements (i.e., Schedule XI) if volumes of manufacture and import do not exceed 10,000 kg/year (50,000 kg accumulated).

Product Development Polymers

Product development polymers are research and development polymers evaluated before full commercialization by using pilot plants, production trials, or customer trials in a program that is two years or less in duration. The intent of this activity is to modify technical specifications in response to performance requirements of potential customers; however, test marketing is not included. This category includes polymers being manufactured on a fee basis for domestic or foreign customers that are conducting research. The Regulations offer the person who is manufacturing or importing product development substances the option of submitting the Product Development Schedule (Schedule XII) in place of the regular notification requirements. There are no volume limitations on product development polymers.

Export-Only Polymers

This category includes polymers imported into or manufactured in Canada and destined solely for foreign markets. At any one time, the quantity of polymer cannot exceed 20,000 kg. Schedule XIII addresses export-only polymers.

Site-Limited Intermediates

An intermediate substance is one that is consumed, in whole or in part, in a chemical reaction used for the intentional manufacture of another substance. A site-limited intermediate polymer is defined in the Regulations as an intermediate not exceeding an accumulated quantity of 50,000 kg; at any one time, the combined inventory of the polymer that is

(a) manufactured for the purpose of being consumed on the site of manufacture,
(b) manufactured at one site in Canada and then transported to a second site in Canada where it is consumed, or
(c) imported and transported directly to one site at which it is consumed does not exceed 20,000 kg.

Annual and Accumulated Volumes

No notification of any news polymer is required if quantities do not exceed 1,000 kg/year of 5,000 kg accumulated. All other polymers are subject to reduced notification requirements if the annual volume of manufacture and import does not exceed 10,000 kg/year and the accumulated volume has not exceeded 50,000 kg.

Table 11.1. Notification periods for polymers

Schedule	Notification period (no. of days before exceeding trigger quantity)
VI	45
VII	45
VIII	90
XI	5
XII	21
XIII	21

A schematic description of the decision scheme for selection of the appropriate information package is presented in Figure 11.1.

Timing of Notification

As noted previously, the CEPA approach to new polymers is based on pre-manufacture, pre-import notification, which requires the submission of specified information to Environment Canada a prescribed number of days prior to exceeding of prescribed volume quantities. The particular notification period for each of the notification Schedules is outlined in Table 11.1. The assessment period (i.e., the number of days in which the Departments must determine any control action is necessary) is a corresponding number of days.

INFORMATION REQUIRED BY REGULATION

Chemical and Physical Property Data

The types of information required in each of the three major polymer notification Schedules (VI, VII, VIII) are identified in Table 11.2. (Readers are cautioned that, for ease of presentation, there are some minor modifications of the regulatory text used to describe certain data elements.) The tiering of the regulatory structure is readily apparent from this side-by-side comparison. Information requirements, in particular those for test data, increase as the level of potential concern increases. Some details on the more important technical elements follow.

Table 11.2. Data requirements for polymers

Data requirement	Sch VI	Sch VII	Sch VIII
Substance Identification			
CAS or IUPAC name	✓	✓	✓
CAS number	✓	✓	✓
Tradename	✓	✓	✓
Percent of monomer/reactants	✓	✓	✓
Percent of additives/impurities	✓	✓	✓
MSDS	✓	✓	✓
Physical-Chemical Properties			
Number-average molecular weight	✓	✓	✓
Residual constituents with MW < 500 and < 1,000 daltons	✓	✓	✓
Water solubility		✓	✓
n-Octanol solubility or octanol–water partition coefficient		✓	✓
Ultraviolet spectrum			✓
Hydrolysis			✓
Dispersibility in water		✓	✓
Physical state	✓	✓	✓
Toxicological Data			
Acute mammalian toxicity		✓	✓
Skin irritation			✓
Skin sensitization			✓
Repeated dose toxicity			✓
Mutagenicity			✓
Ecotoxicological Data			
Fish acute toxicity		✓	✓
Daphnia acute toxicity		✓	✓
Algal toxicity		✓	✓
Ready biodegradability		✓	✓
Exposure Information			
Specific use information	✓	✓	✓
Manufacture or importation information	✓	✓	✓
Distribution/storage/disposal	✓	✓	✓
Environmental release	✓	✓	✓
Human exposure	✓	✓	✓

Substance Name

Polymer nomenclature incorporates the identity of monomers and reactants
used in the manufacture of the polymer. The name of the polymer may, or
may not (at the discretion of the notifier), include monomers or other
reactants that are either incorporated into the polymer or charged to the
reaction vessel at 2% or less by weight.

Chemical Abstracts Service Registry Number

The most precise Chemical Abstracts Service (CAS) Registry Number available for the substance must be obtained. (CAS will not assign a number if the notifier intends to claim the name as confidential, and in such cases a number cannot be provided.)

Structural Formula/Molecular Formula

For polymers, the structural formula should consist of a simple representative diagram that illustrates the key structural features of the polymer molecule (e.g., types of linkages, functional groups, range, and typical values for the number of repeating units). In addition, the type of polymerization (e.g., graft, block, random) must be indicated.

Monomers and Reactants

Monomers and reactants include compounds such as initiators, cross-linking agents, chain-terminating agents, and chain-transfer agents that are intended to become part of the polymer. Monomers or reactants, either incorporated into the polymer or charged to the reaction vessel at 2% or less by weight in the manaufacture of the polymer, also must be reported even if they were not included in the name of the polymer.

Number-Average Molecular Weight

The number-average molecular weight must be determined on the composition having the lowest average molecular weight of any composition intended for import or manufacture. Gel permeation chromatography (GPC) is the method most widely used to determine polymer molecular weights.

Residual Constituents With Molecular Weights of Less Than 500 Daltons and Less Than 1,000 Daltons

The percentage of residual constituents must be determined on the composition that has the lowest average molecular weight of any composition intended for import or manufacture.

Water Solubility

Water solubility at pH 7 is required. This requirement is not applicable if the solubility in water can be shown to be less than 10 mg/L. For polymers subject to Schedules VII or VIII, solubility at pH 1 and 10 also is required, unless the solubility at these pH values is less than 50 mg/L.

Dispersibility in Water

The degree of dispersibility need not be determined; however, if the polymer is intended to be formulated for dispersal in water, this condition must be stated.

Hydrolysis as a Function of pH

Hydrolysis as a function of pH is required only for polymers that are subject to Schedule VIII and have a solubility in water of more than 50 mg/L. The identity of any known hydrolysis products also must be provided.

Ultraviolet-Visible Spectrum

An ultraviolet–visible spectrum that can be used to determine the potential for photodegradation of the polymer is required. Therefore, the wavelength range from 290 nm to 700 nm must be covered.

Solubility in n-Octanol

Solubility in *n*-octanol is not required if the octanol solubility is less than 50 mg/L. If the polymer is subject to Schedule VII or VIII, either the octanol–water partition coefficient or the octanol solubility can be provided.

Toxicological Data

Acute Mammalian Toxicity

Test animals must be dosed by using the same route of exposure that is anticipated to be the most significant route for potential human exposure. The most significant route of potential human exposure for these Regulations refers to exposure of the general population in Canada, which may be different from exposure for workers in an occupational setting.

Skin Irritation and Skin Sensitization

In most cases, properly conducted human patch tests (positive or negative response) are an acceptable alternative to animal testing for skin irritation or skin sensitization. Human-use experience may also be an acceptable alternative to the prescribed test protocols for toxicological endpoints, especially skin irritation or skin sensitization tests (positive response only).

Repeated Dose Mammalian Toxicity

A test report from a study of at least 28 days' duration must be submitted unless a 14-day test was performed before July 1, 1994. In that case, data from the 14-day test are acceptable. Test animals must be dosed by using the most significant route of potential exposure for the general population in Canada.

Mutagenicity

The mutagenicity test requirements in Schedule VIII consist of: an in vitro test for gene mutation; an in vitro test for chromosomal aberrations in mammalian cells; and an in vivo mammalian test for gene mutation, chromosomal aberrations, or another indicator of mutagenicity that permits an assessment of mutagenicity acceptable to the Departments. Some flexibility is given in the choice of in vivo test to permit the most appropriate test to be chosen for the substance. The choice in in vivo test should be based on results from in vitro mutagenicity tests, the structure and the mechanism of action of the substance, and developments in the field of genotoxicity.

An adequate in vivo mutagenicity test must include evidence that the tissue investigated was exposed to the substance or its metabolites.

Ecotoxicological Data

Fish and Daphnia Acute Toxicity

For polymers subject to Schedules VII or VIII, data from either a fish or a Daphnia test is required on the water-soluble portion of the polymer if the water solubility of the polymer is greater than 10 mg/L. However, if the polymer is expected to be cationic in an aquatic environment, data are required from both fish and Daphnia acute toxicity tests performed on the entire polymeric substance.

Algal Acute Toxicity

An algal acute toxicity test, performed on a regular growth medium, is required for all anionic polymers subject to Schedules VII or VIII. Because the toxicity of polycarboxylic acids may be mitigated by the presence of Ca^{2+} or Mg^{2+} ions, algal toxicity of polycarboxylic acids also must be conducted by using a modified algal growth medium (Ca, or Ca and Mg, added to attain a measured hardness of 150.0 mg/L as $CaCO_3$). For polycarboxylic acids used as scale inhibitors, an algal toxicity test must, in addition, be conducted under a third condition (the addition of an equivalent of Ca^{2+} ion to the test compound stock solution).

Biodegradation

For polymers subject to Schedule VIII, biodegradation information is required for the water-soluble portion of the polymer if the water solubilty of the polymer is greater than 50 mg/L. The biodegradability requirement in Schedule VII is limited to cationic polymers. The identity of any known products of biodegradation must be provided.

Test Procedures

The NSN Regulations require that the conditions and the test procedures used for development and reporting of test data be consistent with the conditions and the test procedures of the Organisation for Economic Co-operation and Development (OECD) *Guidelines for Testing of Chemicals* that are current at the time of testing. As the OECD has yet to publish test procedures exclusively for polymers, the appropriateness of an OECD method for these substances must be determined, and any necessary modification should be made (including the use of an alternative method) to ensure the acceptability of test data. The OECD test guidelines are not designed to serve as rigid test procedures appropriate for all substances; rather, they allow flexibility for expert judgment and adjustments to new developments. Credible alternative procedures are considered, consistent with the spirit of the OECD guidelines. In some instances, such as number-average molecular weight, there is not a finalized OECD method for comparison.

The Health Protection Branch (HPB) mutagenicity test guidelines should be regarded as the standard methods for developing mutagenicity test data for new substance notifications. The HPB guidelines are functionally very similar to the equivalent OECD mutagenicity test guidelines; however, they provide additional advice, or different guidance, on the conduct of some tests. The OECD or other mutagenicity test guidelines will be acceptable when, in the opinion of the Departments, they are equally or better suited to measure the mutagenic potential of the substance under investigation, compared to the HPB guidelines.

Alternative protocols include other domestic or internationally recognized protocols, such as test methods developed by the Departments of

Environment or Health, the International Standards Organization (ISO), the American Society for Testing and Materials (ASTM), the U.S. Federal Insecticide, Fungicide, and Rodenticide Act (FIFRA), and the U.S. Toxic Substances Control Act (TSCA). In addition, protocols developed by individual companies or associations may be acceptable. The method used must be clearly referenced and described in sufficient detail to permit evaluation.

STRUCTURE–ACTIVITY RELATIONSHIPS

Relationships exist between the structure of a substance and its physical properties and toxicity. Knowledge of these relationships, particularly within certain chemical groups, can be used to predict the physical, chemical, toxicological, and ecotoxicological properties of a substance.

Qualitative Structure–Activity Relationships

Qualitative structure–activity relationships, referred to as "read-across," provide a qualitative estimate of a particular property and are derived from experimental data on a reference substance or substances (substances with a chemical structure closely related to that of the new substance).

The validity of read-across estimates will largely depend on the structural similarity between the notified substance and the reference substance(s). Read-across estimates are thus applicable where: (a) the notified substance possesses a "trivial" structural difference from the reference substance(s); or (b) the structural difference between the notified substance and the reference substance(s) is not considered "trivial" but will affect the property in a manner that can be accurately predicted. A trivial structural difference between two substances is any change from the notified substance that is not reasonably expected to markedly alter the physical-chemical, biological, or toxicological properties of the substance. However, trivial structural changes would generally not include such differences as: (a) change, modification, introduction, or removal of functional groups or multiple bonds; or (b) positional or geometric isomers.

Quantitative Structure–Activity Relationship Estimates

Quantitative structure–activity relationships provide quantitative estimates of particular properties, and are usually generated by computer programs that use either regression analysis or molecular descriptors that mathematically represent the structural components of a molecule. Linear or multiple regression of a particular property against another property (e.g., octanol–water partition coefficient versus water solubility or vapor pressure versus boiling point) can be used to derive an empirical relationship for one or several classes of chemicals. An estimate calculated by using molecular descriptors can be based either on experimental values for each molecular descriptor or on experimental values for several molecules containing a common molecular descriptor.

All QSAR estimates must be validated by determining: whether all of the structural features of the new substance are represented in the equation, or by the substances used to generate the estimate; whether the estimate is reasonable in comparison with measured data on structurally similar substances; and whether these substances contain any structures that may invalidate the estimate.

Information to support the acceptance of data based on QSARs should include: (a) a validation of the estimate using the recommended standard procedures for the model (this may include printouts of chemicals and/or structures used to generate the estimate and the experimental data for these chemicals); and (b) the level of confidence associated with the estimate.

Other Calculation Methods

Other methods used to calculate data for a notification (e.g., extrapolation of data generated at different temperatures to provide a value at ambient temperature) will be accepted on a case-by-case basis.

WAIVER OF INFORMATION REQUIREMENTS

Under Subsection 26(4) of the Canadian Environmental Protection Act, a request to waive the requirement for any of the prescribed information may be made to the Department of the Environment. The decision to grant a waiver will be made on a case-by-case basis, based on whether at least one of three criteria have been met. The statutory criteria for a waiver of information that are identified in Subsection 26(4) of CEPA are:

(a) in the opinion of the Ministers, the information is not needed in order to determine whether the substance is toxic;

(b) a substance is to be used for a prescribed purpose or manufactured at a location where, in the opinion of the Ministers, the person requesting the waiver is able to contain the substance so as to satisfactorily protect the environment and human life;

(c) it is not, in the opinion of the Ministers, practicable or feasible to obtain the test data necessary to generate the information.

Waiver requests must be submitted in writing as part of a notification package and should include a well-documented rationale to support the request. If the government has granted a waiver of information, then the particulars of the waiver will be published in the *Canada Gazette* in accordance with Subsection 26(5) of CEPA. The waiver notice will contain only (a) the name of the person (or company) to whom the waiver is granted and (b) the type of information to which it relates (e.g., Company X, biodegradability information). The notice will not specify the substance to which the waiver applies.

ASSESSMENT FOR TOXICITY

The purpose of the New Substance Notification assessment process is to determine whether or not the substance is, or is suspected of being, "toxic" as defined within CEPA. Consequently, the determination of whether a polymer is, or is suspected of being, toxic involves assessment of the potential for exposure to humans and components of the environment, and of the adverse effects of the substance on humans or the environment (including other living organisms, interacting natural systems, and the abiotic components of the environment).

The potential for exposure to a substance depends on the quantity, rate, frequency, and conditions of release of the substance into the environment at all points in its life-cycle, as well as the mobility, environmental compartmentalization, and persistence of the substance. The exposure assessment considers the use of the substance identified by the notifier, as well as other possible ways that the substance might be used if it were placed on the DSL without restrictions.

The assessment of adverse effects on humans and other living organisms considers endpoints such as lethality, mutagenicity, reproductive effects, and organ toxicity, whereas adverse effects on the abiotic components of the environment include consequences such as depletion of the ozone layer, global warming, and production of acid rain.

ACTION TAKEN AFTER AN ASSESSMENT

After the assessment, if there is no suspicion that the substance is toxic, the notifier may proceed with import or manufacture after the assessment period has expired.

When the government suspects that a substance may be toxic, the following measures under Section 29 of CEPA may be taken:

(a) Permit the manufacture or import of the substance subject to specified conditions;

(b) Prohibit the manufacture or import of the substance for a period not exceeding two years (this prohibition lapses at the end of the two-year period unless, before the end of this period, a notice of proposed regulation under Section 34 of CEPA is published in the *Canada Gazette*); or

(c) Prohibit the manufacture or import of the substance until supplementary information or test results have been submitted to the government and assessed (the assessment period for this supplementary information expires 90 days after receipt of the information, or at the end of the original assessment period, whichever is the later date).

Any of the above actions must be taken before the expiration of the assessment period. The notifier must comply with these measures or withdraw the notification.

When a condition or a prohibition is issued or altered, a notice must be published in the *Canada Gazette* describing the action and the substance to which it applies. The name of the notifier is not included in this notice. Furthermore, if publication of the substance name would result in the release of confidential business information, a masked name is published.

Polymers that have been assessed and are not anticipated to pose a risk to Canadians are eligible to be placed on the DSL. They will be added to the List when the notifier advises that regulatory specified listing volumes have been exceeded.

REFERENCE

SSC. 1993. *Guidelines for the Notification and Testing of New Substances: Chemicals and Polymers: Pursuant to the New Substances Notification Regulations of the Canadian Environmental Protection Act.* Ottawa, Ontario, Canada: Supply and Services Canada, Catalogue No. En. 49-2912-1993E.

Chapter 12

Regulation of New Polymers in the Pacific Region — Part I

■ GORDON REIDY
  Woodward-Clyde Pty. Limited

INTRODUCTION

In the past few years there has been a substantial increase in the number of Pacific region countries developing national schemes for the notification and assessment of industrial chemicals. At present two countries in the Pacific region functionally differentiate polymers from nonpolymeric substances — Australia and South Korea. Of these two countries, only Australia has developed criteria to assess the environmental hazard of polymers. This review will focus on national assessment schemes in the Pacific region that functionally distinguish polymers from nonpolymers and will also provide information on the nature of other schemes and how they assess polymers. It should be noted that the environmental regulations on polymers that have been implemented by the Japanese government are covered in Chapter 13.

SCHEMES THAT DISTINGUISH POLYMERS AND NONPOLYMERS

Australia

Overview

The National Industrial Chemicals Notification and Assessment Scheme (NICNAS) came into effect on July 17, 1990. The Scheme is responsible for assessing both new and existing industrial chemicals. A new chemical is one that is not listed on the Australian Inventory of Chemical Substances (AICS). An existing chemical is listed on the AICS. New chemicals are assessed for their human health and environmental effects prior to their import or manufacture in Australia. Assessment takes 50 to 90 days. At the end of the assessment period a report is published that outlines the environmental and human health hazards of the new chemical and the risks the use of the chemical may pose to workers, the public, and the environment. Existing chemicals are assessed for their human health and environmental effects under the Existing Chemical Scheme. Notice of the assessment of an Existing Chemical is given in the *Chemical Gazette*, which is published by the Commonwealth Government the first Tuesday of each month. The data required for the assessment of an existing chemical are detailed by the Director of NICNAS in the *Chemical Gazette*. A chemical may become a Priority Existing Chemical following nomination by the public, industry, or other groups as a result of immediate concerns surrounding its use.

New chemical assessment under NICNAS is modeled on the European Union's notification schemes and requires the provision of data based on the OECD Minimum Premarketing Data Set (MPD). Under NICNAS new synthetic polymers (as opposed to biopolymers) are considered a distinct chemical class and have data requirements that differ from those of non-polymeric substances. At present a polymer is defined as "a chemical comprising a simple weight majority of molecules containing 2 or more monomer units which are covalently bound to at least 2 other monomer units and for which the number average molecular weight is more than 500." This definition will be changed in the near future to harmonize with the OCED definition of a polymer.

A new synthetic polymer is one that contains a combination of monomers or other reactive components not listed in the AICS. Here, the 2% rule applies. To make a polymer "a new synthetic polymer," a new monomer must be incorporated into an existing polymer at a final concentration greater than 2%w/w. This is similar to the position under the U.S. Toxic Substances Control Act (TSCA), the difference being that the Australian legislation applies the 2% rule to the finished polymer on an as-incorporated basis, whereas the U.S. TSCA regulations allow industry the option to apply the 2% rule on as-charged or an as-incorporated basis.

The number-average molecular weight (NAMW) of the polymer determines the data required for assessment under NICNAS. If the NAMW of the new polymer is greater than 1,000 daltons, it is a priori assumed that the polymer will be less hazardous to the environment than it would be if the NAMW were less than 1,000; and so the data requirements for assessment are reduced. If a polymer has an NAMW < 1,000, it is assessed in the same manner as a nonpolymeric substance, as shown in Figure 12.1. In effect, only

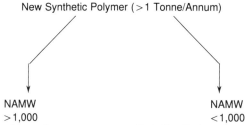

New Synthetic Polymer (>1 Tonne/Annum)

NAMW >1,000 **Reduced Data Requirements**	NAMW <1,000 **Data Not Reduced**
C.A.S. Number	Data Requirements in Left Column, Plus:
Chemical Name	Acute Oral Toxicity
Molecular and Structural Formulae	Acute Dermal Toxicity
Molecular Weight	Acute Inhalation Toxicity
Spectral Data	Skin Irritation
Purity and Hazardous Ingredients	Eye Irritation
Use	Skin Sensitization
Appearance	Bacterial Mutation
Quantity	Clastogenicity
Occupational Exposure	10–28 Day Repeated Dose Toxicity
Environmental Exposure	Fish Acute Toxicity
Public Exposure	*Daphnia* Acute Immobilization and
Melting Point	Reproduction Test
Density	Algal Growth Inhibition
Vapour Pressure	Ready Biodegradability
Water Solubility	Bioaccumulation
Hydrolysis	
Partition Coefficient	
Adsorption/Desorption	
Dissociation Constant	
Particle Size	
Flash Point	
Flammability	
Autoignition	
Stability	
Methods of Identification	
Material Safety Data Sheet	
Label	
Percent Monomeric Composition	
Residual Monomers	
Low Molecular Weight Species	
Degradation Products	

Figure 12.1. Data requirements under Australia's National Industrial Chemicals Notification and Assessment Scheme.

polymers with an NAMW > 1,000 are assessed differently from other chemicals.

Assessment of Environmental Effects

The assessment of the environmental effects of polymers with NAMW > 1,000 under NICNAS is divided into two parts, as outlined below:

1. *Hazard Identification.* The following five physical-chemical parameters are used in assessing the environmental hazard of a polymer:
   (a) *Molecular weight.* The NAMW and the percentage of low molecular species or oligomers are critical parameters in determining the environmental hazard of a polymer. As the molecular weight of a polymer increases, so does its size. It is generally accepted that as the molecular weight of an organic compound approaches 1,000, absorption across biological membranes tapers off to zero (Grandjean, 1990; Klaassen and Rozman, 1991). There are examples of special transport mechanisms in cells for the movement of large molecules such as proteins, but these mechanisms are highly selective and generally do not affect the absorption of foreign compounds such as synthetic polymers. If a polymer cannot be absorbed because of its high molecular weight, then it cannot cause systemic toxicity in animals or plants in the environment. Polymers of high molecular weight (i.e., > 1,000 NAMW) thus are regarded as presenting a low hazard to the environment and do not require the provision of ecotoxicity data for assessment. The concentration of low molecular weight species is also considered in determining the environmental hazard of a polymer. A polymer that has a small proportion of low molecular weight species (i.e., those with < 1,000 NAMW) is considered less hazardous than a polymer with a higher proportion of low molecular weight species. This is so because low molecular weight species (especially those below 500 NAMW) may be readily absorbed across biological membranes and thus cause systemic toxicity in plants and animals.
   (b) *Cationic charge density.* Cationic polymers containing functionalities such as quaternary nitrogen may cause nonsystemic toxicity to aquatic organisms such as fish and algae by interacting with exposed biological membranes such as gills and cell walls. If a polymer has a cationic charge density less than 5,000 then it is considered hazardous if it is dispersible in water. The charge density figure of 5,000 is derived from work undertaken by the

U.S. Environmental Protection Agency (U.S. EPA, 1993), which shows that polymers with a cationic charge density of 5,000 or greater do not exhibit significant aquatic toxicity.

(c) *Water solubility.* The water solubility of a polymer determines how widely the polymer may be distributed in the aquatic environment and its concentration. The more water-soluble a polymer, the more widely it may be distributed and the higher its potential concentration in water. In Australia, where receiving waters are often of low flow, this means that the concentration of a water-soluble polymer may reach levels high enough to cause toxicity to aquatic organisms. Thus, the greater the water solubility of a polymer, the more hazardous it may be to the environment.

(d) *Dissociation constant.* Polymers that contain basic nitrogen groups that are likely to be protonated in the environmental pH range of 4 to 9 to give cationic species are considered hazardous. The environmental hazards of these polymers are as described in item (b) above.

(e) *Stability/Degradation.* Polymers that readily degrade or de-polymerize, either during use or after disposal, are likely to pose a hazard to the environment. This is so because such polymers may break down either to small molecular species ($< 1,000$ NAMW) or to monomeric constituents. The low molecular weight (and hence size) of these species means that the degradation products may be absorbed by plants or animals and thus be able to cause local and/or systemic toxicity.

2. *Exposure Assessment.* Information on the amount of a polymer released to the environment during manufacture or use is critical for the assessment of its environmental effects. Under NICNAS, environmental release means release to the environment by venting, wastewater streams, or sewer, not land disposal. If release to the environment is small or occurs at low concentrations, then the risk to the environment is also small. If there is no release, then there is no risk to the environment, no matter how hazardous the polymer. If release to the environment is large or occurs at high concentrations, then it is of concern. Of particular concern in Australia is release of polymers to the aquatic environment. Receiving waters from industrial processes are generally of low flow and small volume; so the concentration of a polymer in the aquatic environment may reach relatively high levels. Estimation of the concentration of polymers in receiving waters or waste streams thus is important in determining whether aquatic systems may be at risk.

The determination of the environmental effects of a polymer follows from the hazard and exposure assessment. If both the hazard and the environmental concentration of the polymer are high, then recommendations may be made to reduce release or to mitigate the effects of release. If the hazard and the environmental release of a polymer are low, then the polymer may fit within the definition of a polymer of low concern.

Environmental Criteria for Polymers of Low Concern

Under NICNAS, a polymer is considered of low concern to the environment if it satisfies all of the following criteria:

- NAMW is greater than 1,000 with less than 2% (w/w) of the polymer having a molecular weight less than 500 and less than 5% (w/w) of the polymer having a molecular weight less than 1,000.
- It is not a polycarboxylic, polyaromatic sulfonate, polyaliphatic sulfonate, or polycationic polymer; or if it is a polycationic polymer then it does not have a charge density of less than 5,000.
- It does not contain functional groups likely to give cationic or anionic species in the pH range 4 to 9.
- It has a water solubility < 1 mg/L.
- It is stable under use conditions and does not readily degrade or depolymerize.

A "new" polymer satisfying these criteria is a priori considered to be of low environmental concern and may be introduced into Australia in a shorter period of time and with fewer data than required for higher concern polymers.

South Korea

The Toxic Chemicals Control Law (TCCL) came into effect on February 8, 1991. The Law regulates the import and the manufacture of industrial chemicals in Korea. If a chemical is not listed on the Existing Chemical List, is not regulated by other legislation, is not an excluded chemical, is not used solely for research and testing, and is not imported in quantities of 100 kg/year or less, then it must be notified to the Ministry of Environment for assessment. Data requirements under the scheme are reduced for chemicals (polymers) that satisfy all of the following criteria:

- Contain sequences that are built up from one or more types of monomer unit.

- Show a distribution of molecular weights primarily due to differences in the number of monomer units.
- Contain at least three monomer units covalently linked to at least one other monomer unit or another reaction component.
- Contain at least 50% by weight of the species described above but no more than 50% of species with the same molecular weight.

Polymers satisfying these criteria are assessed by using only the following physical-chemical parameters:

- Number-average molecular weight.
- Residual monomer concentration and concentration of species with less than 1,000 molecular weight.
- Monomeric composition of the polymer.
- Solubility in water and common organic solvents such as methanol, *n*-octanol, dimethylformamide.
- Melting point, structure, and stability in acidic and basic solutions.
- Use information.

SCHEMES THAT DO NOT DISTINGUISH POLYMERS AND NONPOLYMERS

New Zealand

Under the Hazardous Substances and New Organisms Bill (HSNO) industrial, agricultural, and veterinary products will require a hazard assessment before introduction (by import or manufacture) into New Zealand. If a chemical is determined to be hazardous, a submission must be made to the New Zealand Government before import or manufacture occurs. The hazard classification system is based in part on the United Nations recommendations for the Transport of Dangerous Goods. The Bill will apply not only to new chemical entities but also to products. Polymers are not differentiated from other chemicals in the Bill or its proposed regulations. The Bill is at present undergoing review by a parliamentary committee following a period of public comment.

The Phillippines

Republic Act 6969 (1990) establishes a mechanism for control of the import and the manufacture of industrial chemicals. The Act establishes a Phillippine Inventory of Chemicals and Chemical Substances. A chemical substance that is not listed on the Inventory must first be notified to and assessed by the Phillippine Government before it can be manufactured or imported. The Act does not differentiate polymers from other chemical substances. However, a guidance document issued by the Department of Environment and Natural Resources in May 1995 suggests that polymers may be subject to reduced data requirements when final regulations are issued.

Malaysia

The Draft Industrial Chemical Act (1995) regulates the import and the manufacture of industrial chemicals in Malaysia. The Act establishes the Malaysia Inventory of Industrial Chemicals. A chemical not listed on the Inventory is considered new and must be assessed by the Malaysian Government before sale. The Draft Act does not distinguish polymers from other chemical substances.

China

The rules and regulations for the "First Chemical Import and Toxic Chemical Import and Export Environment Stipulation" provide a mechanism for the control of industrial chemicals going into and out of China. The rules and regulations came into effect on February 26, 1995. Although the rules and regulations identify polymers as a separate chemical class, polymers apparently are assessed identically to other industrial chemicals, as the data required for assessment of all chemicals depend on import volume rather than chemical nature.

Other Pacific Rim Countries

At the time of writing of this chapter, national schemes for the assessment of industrial chemicals in Taiwan, Indonesia, and Thailand are in development or do not exist.

REFERENCES

Chemical Registration Centre of the National Environment Protection Agency. 1995. *Registration of First Time Chemical Import and Toxic Chemical Import and Export Control Rules and Regulations.* People's Republic of China.

Department of Environment. 1995. *Draft Industrial Chemicals Act.* Department of Environment, Malaysia.

Department of Environment and Natural Resources. 1990. *Rules and Regulations of Republic Act* 6969. Republic of the Phillippines.

Grandjean, P. 1990. *Skin Penetration: Hazardous Chemicals at Work.* London: Taylor and Francis.

Klaassen, C. D. and K. Rozman. 1991. Absorption, distribution and excretion of toxicants, in *Casarett and Doull's Toxicology*, 4th Ed., ed. M. O. Amdur, J. Doull, and C. D. Klaassen. New York: McGraw-Hill.

Ministry for the Environment. 1994. *Proposal for Regulations Under the Hazardous Substances and New Organisms Bill.* Wellington, New Zealand.

Ministry of Environment. 1993. *The Toxic Chemicals Control Law, A Guide for the Chemical Manufacturers and Importers.* Seoul, Korea: Ministry of Environment.

National Industrial Chemicals Notifications and Assessment Scheme. 1990–1995. *Handbook for Notifiers.* Commonwealth of Australia.

U.S. EPA. 1993. Premanufacture notification; revisions for notification regulations, Parts 700, 720, 721, and 723, in *Federal Register,* pp. 7679–7701. Washington, DC: U.S. Government Printing Office.

Chapter *13*

Regulation of New Polymers in the Pacific Region—Part II (Japan)

■ RONALD L. KEENER
Rohm and Haas Company

INTRODUCTION

Laws and Regulatory Authorities

In 1973, Japan became the first country in the world to develop a chemical control law to control general industrial and commercial chemicals. The law, which is commonly referred to as the Chemical Substances Control Law (CSCL), had as its primary purpose to protect human health from the adverse effects of chemicals by environmental exposure and, particularly, via the food chain (Japan, 1973, and revisions). The law was substantially modified in 1986 to reflect new knowledge and to attempt harmonization with similar laws in other countries (Japan, 1986, and revisions). The law is administered primarily by the Ministry of International Trade and Industry (MITI) in association with the Ministry of Health and Welfare (MHW) and the Environmental Agency (EA).

The overall regulatory scheme used by Japan distinguishes polymers from nonpolymers. Regulatory schemes used by other Pacific region countries are described in Chapter 12 of this book.

Regulatory Approach under the CSCL

The control of industrial chemicals in Japan under the CSCL is augmented by the Industrial Safety and Health Law (ISHL). This law, administered by

265

the Ministry of Labor, was issued originally in 1972 (Japan, 1972) and later amended in 1979. The primary purpose of this law is worker protection, with a major focus on protection against carcinogenic substances.

The primary means employed in the CSCL for the control of chemical substances was the initial development of an inventory of chemical substances currently in commerce in Japan, with a requirement for industry subsequently to notify new chemical substances (i.e., substances not on the initial inventory) 90 days prior to their introduction (manufacture or import) into Japan. A safety assessment of the newly notified chemical is carried out by government authorities during this 90-day period. If the chemical is deemed safe, the notifier is issued an approval certificate at the end of the 90 days, which allows the notifier to freely manufacture or import the new chemical without regulatory controls. Substances deemed not to be safe are subject to specific regulatory controls before they can be imported or manufactured after the 90-day review period. Newly notified chemicals eventually are added to the Japanese inventory of chemical substances. In addition to requirements for notification and control of new chemical substances, the CSCL contains provisions for assessment and control of existing chemicals.

The assessment of the safety of chemicals in Japan is based upon three major properties: biodegradability, bioaccumulation, and chronic toxicity. Substances that are biodegradable to harmless substances are generally regarded as "safe chemicals." Substances that are not biodegradable but possess both bioaccumulation and chronic toxicity properties are classified as Class I Specified Chemicals. Substances that are both nonbiodegradable and nonbioaccumulative but possess potential chronic toxicity properties are classified as Designated Chemicals. Those Designated Chemicals that have high environmental exposure and are demonstrated to have chronic toxicity properties are classified as Class II Specified Chemicals. Chemicals that are demonstrated not to possess chronic toxicity properties are considered to be safe chemicals regardless of their biodegradation or bioaccumulation properties. The severity of regulatory controls decreases in the following order: Class I > Class II > Designated Chemicals. Safe chemicals are not subject to any regulatory controls under the CSCL.

SUBSTANCES SUBJECT TO NEW CHEMICAL NOTIFICATIONS

General Requirements

Any chemical substance that is not listed on the Japanese inventory of chemical substances or that is not exempt from listing on the inventory (as described below) is subject to the new chemical notification requirements of the CSCL before it can be manufactured or imported into Japan. A newly notified chemical is not actually added to the Japanese inventory until at least 18 months after the date of the initial notification. Consequently, any other company wishing to manufacture or import the same new chemical during the delayed listing period also must submit a separate new chemical notification. However, customers of the notifier may be able to import the new chemical during the delayed listing period if they can present a copy of the original submitter's Certification of Approval at the point of importation.

The Japanese Chemical Substances Inventory

The CSCL required MITI to maintain a published inventory of the initial existing chemical substances in commerce when the laws was passed and of all new chemical substances notified and approved under the new chemical notification process. MITI meets the requirement by publication of the *Handbook of Existing and New Chemical Substances*, commonly called the *MITI Handbook*. The lastest edition (Seventh Edition) of the handbook was published by the Chemical Daily Co., Ltd. (Japan, 1996). Existing and new chemicals appear in separate sections of the handbook. In addition to listed chemicals, the handbook also contains pertinent sections of the law and implementing regulations, as well as guidelines for conducting tests to be submitted with new chemical notifications.

Unlike the case of other countries with chemical control laws and inventories, chemicals are not listed on the MITI inventory by Chemical Abstract Registry Numbers (CASRN) and names; instead they are listed by special numbers (commonly called MITI numbers) and names unique to Japan. In the existing chemicals section of the *MITI Handbook*, MITI numbers in some cases correspond to discrete chemical substances, whereas in other cases they correspond to generic classes of chemicals (e.g., alkyl acrylate and alkyl methacrylate copolymer). MITI numbers in the new chemicals section of the *MITI Handbook* normally correspond only to discrete chemical substances. The existing chemicals section of the inventory contains about 20,000 MITI numbers, and the new chemicals section contains about 1,500 listings. The actual number of chemicals covered by these listings, however, is estimated to be in excess of 50,000 as a result of the generic listings.

The rules for defining polymers in Japan for the purpose of inventory listing are also different from those of other countries that require polymers to be inventory-listed. In most of these other countries, monomers and other polymer reactants present at concentrations of less than 2% in the polymer can be ignored for purposes of defining the polymer. In Japan, polymers can be defined in terms of the three top monomers present in the polymer; all others can be ignored unless they impart special functionality to it. However, block and graft polymers do not have to be listed on the inventory if the individual block or graft sections of the polymer are already listed on the inventory. Ammonium and amine salts of anionic polymers also do not require listing if both the amine and free acid forms of the polymer are listed on the inventory.

Exemptions

Exemptions Not Requiring Notification

The following substances do not require any notification under the CSCL:

1. Chemicals regulated under other laws:
   (a) Pharmaceuticals and their intermediates.
   (b) Food additives, containers, and wrappings.
   (c) Agricultural chemicals, fertilizers.
2. Site-limited intermediates.
3. Research and development (R&D) chemicals.
4. Impurities present at less than 1%.
5. Articles.
6. Formulated mixtures (although individual components must be listed).
7. Certain finished products intended for use by the consumer (e.g., paints).

Special forms must be completed for customs import of pharmaceutical intermediates as well as research and development chemicals. The site-limited intermediate exemption was modified in 1984 to allow transfer between sites controlled by the same manufacturer.

Exemptions Requiring Limited Notification

The CSCL allows small quantities of new commercial chemicals to be manufactured and/or imported without a full new chemical notification as long as the cumulative total for all manufacturers/importers does not exceed one ton per year. Preapproval before manufacture or import is required. The exception must be renewed annually. Data concerning projected volumes, uses, and chemical identity are required.

NOTIFICATION AND ASSESSMENT OF NEW SUBSTANCES

General Considerations

Japan has separate assessment schemes and data requirements for polymeric and nonpolymeric substances as described below. Polymers that do not meet the regulatory definition of polymer and polymers that do not pass the polymer assessment scheme must be assessed under the nonpolymer assessment scheme (Standard Chemicals).

Unlike those of other countries, the assessment schemes do not call for a single defined set of data to be submitted for all new chemicals. Certain basic types of data are required for all substances, but the need for further tests is based on a tiered testing approach in which the results of a specific test will determine whether the next tier of tests is required.

Persons wishing to manufacture or import a new chemical must submit the notification and required data 90 days before intended import or manufacture. The authorities will issue a Certificate of Approval for substances assessed to be safe at the end of the 90-day review period, and they will prescribe specific control measures for substances deemed not safe (i.e., Class I, Class II, and Designated Chemicals).

After a notification has been approved, the new chemical substance will not be added to the official inventory of new chemical substances for at least 18 months. Until the official listing is announced in the *Government Gazette*, no one other than the original submitter and certain of the submitter's customers may import or manufacture the new substance without submitting a new notification package.

Standard Chemicals

Assessment Scheme/Data Requirements

The *basic data* to be submitted for all standard chemicals include the following:

- Chemical identity data, including impurities, uses, and volumes.
- Physicochemical data corresponding to the OECD minimum premarket data (MPD) set as shown in Table 13.1.

In addition to the basic data above, biodegradation tests and selected additional tests from the following tiered testing schedules may be required, depending on the results obtained from the preceding tiered testing schedule:

- Biodegradation tests (OECD 301C and 302C) (Tier 1).
- Bioaccumulation tests (Tier 2).
- Screening toxicity tests (Tier 3).
- Chronic toxicity tests (Tier 4).

Table 13.1. OECD physicochemical test requirements for standard chemicals

■ Melting point	■ Spectra (UV/IR)
■ Boiling point	■ Solubility data
■ Specific gravity	—Water
■ Vapor pressure	—Chloroform
■ Octanol/water partition coefficient	—Methanol
■ Hydrolysis	—*n*-Hexane
■ Adsorption/desorption	—Ethyl acetate
■ Dissociation constant	—Tetrahydrofuran
■ Particle size distribution	—Dimethyl sulfoxide

The detailed tests involved with the above four items are listed in Table 13.2. The *MITI Handbook* contains descriptions of the actual test methods used for these tests. In general, all required tests are run according to OECD test methods and according to OECD Good Laboratory Practice (GLP). Nevertheless, there are frequent variations between OECD requirements and Japanese test requirements; so one should specifically review the *MITI Handbook* and discuss test requirements with the authorities before conduc-

Table 13.2. Tiered test requirements for standard chemicals

Tier 1 — Biodegradation tests (OECD 301C, 302C)

Tier 2 — Bioaccumulation tests

- Octanol/water partition coefficient (OECD 107)
- Fish bioaccumulation (OECD 305C)

Tier 3 — Screening toxic tests

- Ames mutagenicity test (5 strains) (OECD 471)
- Chromosomal aberration test (OECD 473)
- 28-day oral toxicity study (OECD 407)

Tier 4 — Chronic toxicity tests

- Chronic toxicity test
- Reproductive toxicity test
- Teratogenicity test
- Carcinogenicity test
- Toxicokinetic test
- Pharmacological test

ting tests. In many cases, test results will be accepted only if conducted at a laboratory specifically approved by MITI/MHW. For certain tests such as biodegradation and bioaccumulation, tests must be conducted on specified Japanese sludge and fish species. Therefore, it may be advantageous to have these tests run in Japan.

Figure 13.1 presents a flow diagram showing the Standard Chemical Assessment Scheme and the decision points at which the need for additional testing has to be established. The authorities will decide whether a designated substance requires additional chronic toxicity tests to determine whether it is a Class I Specified Chemical after review of the environmental exposure potential of the chemical.

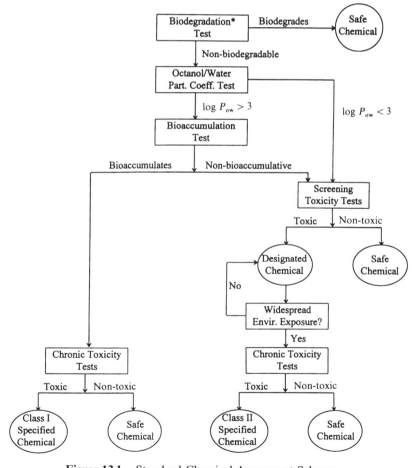

Figure 13.1. Standard Chemical Assessment Scheme.

Polymers

Qualifying Polymers

A polymer must meet the following criteria to be assessed under the Polymer Assessment Scheme described below:

- Have a number-average molecular weight greater than 1,000.
- Have a molecular weight distribution.
- Have standard polymer properties such as an unclear melting point.

Polymers that do not meet these criteria do not qualify for the Polymer Assessment Scheme and must be tested by the Standard Chemical Assessment Scheme. Polymers that qualify for the Polymer Assessment Scheme but do not pass it also must be tested under the Standard Chemical Assessment Scheme.

Polymer Assessment Scheme/Data Requirements

The *basic data* to be submitted with a new polymer notification are similar to those for a standard notification and include:

- Chemical identity data.
- Composition and impurities.
- Uses.
- Three-year production estimates.

The following *additional data* elements are required for a new polymer notification (note that all tests must be run and a favorable outcome obtained for a polymer to be classified as safe):

- Biodegradation test (if practical).
- Chemical stability tests.
- Solubility data in water and solvents.
- Judgment of structural characteristics. For insoluble polymers, data must be provided that confirm that the polymer is totally cross-linked or crystalline. The data may include information on chemicals used in the manufacture of the polymer.
- Molecular weight distribution. Information provided should be sufficient to confirm that the polymers contain less than 1% by weight of polymer with a molecular weight of less than 1,000. GPC is the preferred method.

Table 13.1 summarizes the data requirements of the above five items, and a schematic diagram of the Polymer Assessment Scheme appears in Figure 13.2. The authorities decide whether or not the polymer is safe on the basis of compositional information, comparison of similar and/or contained polymers or functional groups, and known toxicological properties. Polymers containing heavy metals are always deemed unsafe. MITI has not issued a formal list of functional groups in a polymer that might lead to an unfavorable decision during this assessment. Experience suggests that polymers composed only of monomers considered safe will be considered safe in this analysis. However, there is also some evidence that functional

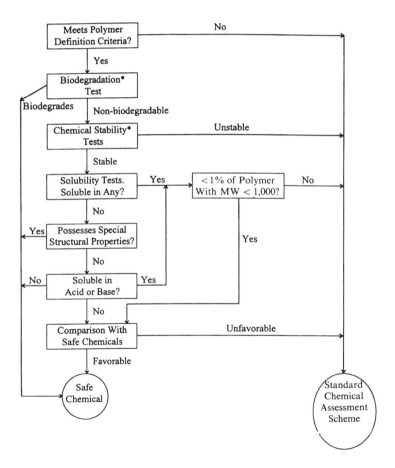

Figure 13.2. Polymer Assessment Scheme. *If the polymer degrades or decomposes to unsafe chemicals or to chemicals of unknown safety, these decomposition products may also be subject to separate safety assessments.

groups that would result in a polymer's not being considered low risk in other countries may result in the polymer's also being considered of concern in Japan.

The Polymer Assessment Scheme for polymers was developed as an alternate to the Standard Chemical Assessment Scheme for a number of reasons:

- Polymers are frequently insoluble in water, and so many of the standard tests are not applicable.
- Many of the OECD physicochemical tests are not applicable to polymers.
- If the polymer backbone can be shown to remain intact after the chemical stability tests, it can be assumed that the polymer will not readily undergo biodegradation.
- Polymers with a molecular weight greater than 1,000 cannot readily penetrate biological membranes, and thus should not bioaccumulate.
- Solvents other than water and octanol are necessary for solubility screening, as these are solvents commonly found in the Japanese environment.
- Tests conducted under acidic conditions are necessary in order to simulate effects in the digestive tract.

CONTROLS ON REGULATED CHEMICALS

Class I Specified Chemicals

Class I Specified Chemicals are chemicals that biodegrade, bioaccumulate, and possess chronic toxicity properties. They are the most stringently regulated chemicals, and in some cases may be subject to total bans. Persons wishing to manufacture or import Class I Specified Chemicals are required to:

- Apply for approval to manufacture or import them.
- Provide periodic reports on manufacture or import volumes.
- Restrict uses of the chemicals to those specified by the authorities.

Class II Specified Chemicals

Class II Specified Chemicals are chemicals that are nonbiodegradable and nonbioaccumulative and possess chronic toxicity properties. Persons wishing to manufacture or import these chemicals must submit annual reports of production or import volumes and are not allowed to exceed production/import limits imposed by the authorities.

Designated Chemicals

Designated Chemicals are those that are nonbiodegradable and nonbioaccumulative and are suspected of possessing chronic toxicity properties (based on toxicity screening test results), but are not widely distributed in the Japanese environment. Manufacturers and importers of these chemicals must submit annual manufacture/import volume reports.

REFERENCES

Japan. 1972. *Industrial Safety and Health Law.* Ministry of Labor, Government of Japan.

Japan. 1973. *Law Concerning the Examination and Regulation of Manufacture, etc., of Chemical Substances.* 1973 (Law No. 117), effective April 16, 1974 and amended 1975. Government of Japan.

Japan. 1986. *Law Concerning the Examination and Regulation of Manufacture, etc., of Chemical Substances.* Revision of May 17, 1986, effective April 1, 1987. Government of Japan.

Japan. 1996. *Handbook of Existing and New Chemical Substances.* Seventh Ed. Ministry of International Trade and Industry, Chemical Products Division, Government of Japan.

Chapter 14

European Environmental Legislation for Polymers

■ MARIE-PIERRE RABAUD
  Exxon Chemical France

INTRODUCTION

European environmental legislation has grown steadily over the past decade. Enlargement of the European Union to fifteen members, through admission of Austria, Finland, and Sweden in 1995, and the integration of concern for the environment into other policy areas are among the important changes that will have a significant overall impact on the form and the content of future environmental legislation in the European Union.

OVERVIEW OF THE LEGISLATIVE PROCESS IN THE EUROPEAN UNION

The European Union

The drive toward European integration originated with the establishment of the European Economic Community (E.E.C.), set up by the Treaty of Rome in March 1957. The Treaty of Rome followed establishment of the European Coal and Steel Community (ECSC) by the Treaty of Paris in April 1951 and the European Atomic Energy Treaty (EURATOM) of March 1957. In 1987, the Single European Act (S.E.A.) took effect and modified the Treaty of Rome for the first time. The purpose of the S.E.A. was to accelerate the process of the internal market by 1993. After the S.E.A., the next landmark in regulatory development of the European Communities was the proposal of the Treaty on European Union, or, the Maastricht Treaty, in 1991. It was signed at Maastricht in The Netherlands in February 1992 and took effect on January 1, 1993.

The Maastricht Treaty provided for the transformation of the market into a European Union, whose chief characteristics are that it is:

- A political union (e.g., with establishment of a common foreign and security policy, rights of European citizenship, new powers and responsibilities for the European Community, and additional powers to the European Parliament).
- An economic and monetary union.

The period since the signing of the Treaty of Rome has seen more than a doubling in the number of Member States in the European Community; the six founding Member States — Belgium, France, West Germany, Italy, Luxembourg, and the Netherlands — were joined by Denmark, Ireland, and the United Kingdom on January 1, 1973, by Greece on January 1, 1981, and by Spain and Portugal on January 1, 1986. By 1995, Austria, Finland, and Sweden joined the Union. The original Community of six is now a union of fifteen, with other countries from the Mediterranean and Central and Eastern Europe waiting their turn.

The Legislative Process

Within the European Union (E.U.), the European Community (E.C.) is responsible for proposing and adopting legislation that is relevant to the Internal Market. This legislation falls within the scope of the Treaty of Rome. Five bodies (Community Institutions) are involved in the process of developing legislation based on a complex system of relationships. These bodies are:

- The European Commission
- The Councils of Ministers
- The European Parliament
- The European Courts
- The Economic and Social Committee

The Courts are not directly involved in preparation and approval of legislation, but they are responsible for interpreting Community law and issuing judgments in disputes between the Community Institutions, Member States, and/or third parties. Decisions of the Courts have an important influence on creation and development of European law, including European environmental law. The Economic and Social Committee is a tripartite body, representing the various categories of economic and social activity appointed to assist and advise the Council and the Commission. Its opinions are not binding on the Council and the Commission, but can influence them.

Community law is drawn up in a process whereby the Commission proposes and the Council decides after consulting the European Parliament.

The European Commission

The European Commission, which has the sole right to initiate legislation, is often referred to as the "Guardian of the Treaties." It is composed of 20 members or "Commissioners" appointed for a term of five years. There must be at least one and not more than two national representatives of each member state in the Commission. Each Commissioner is supported by one or several Directorates General (DGs), covering all Community activities and responsibilities. There are 24 DGs. The following DGs are most concerned with environmental issues:

- DG III: Industry, covering activities such as harmonization of national measures on emission and product standards, as well as classification and labeling of dangerous preparations.
- DG XI: Environment, Nuclear Safety and Civil Protection, concerned with issues such as air, water, and soil protection, waste management, industrial installations and their emissions, environmental control of industrial products, and biotechnology.

The Commission issues a program at the beginning of its five-year term of office that broadly outlines actions and legislation it intends to implement. A specific program is issued each year. Moreover, it may issue programs precisely covering certain areas, such as the Fifth Environmental Action Programme of February 1, 1993, which set the pace for the Community's policies and actions on environmental matters toward the end of the century. The overall goal of this program is "Towards Sustainability."

The Councils of Ministers

There are numerous Councils, which cover particular sectors of interest (e.g., energy, environment, internal market, and consumers). These Councils are formed by the responsible ministers from each government (i.e., 15 members for each Council), and the Commissioner responsible for the subject of interest attends Council meetings.

The Council Presidency is filled in successive six-month periods by each Member State in a rotation roughly based on alphabetical ordering of the Member States, as spelled in their national tongue — currently as follows: Belgium, Denmark, Germany, Greece, Spain, France, Ireland, Italy, Luxembourg, The Netherlands, Austria, Portugal, Finland, Sweden, and the United Kingdom. The order of rotation of countries was altered in 1993 in order to avoid one country's presiding over the Council during the same half-year as its prior term. The Council presidencies have been scheduled as follows: 1996, Italy and Ireland; 1997, The Netherlands and Luxembourg; 1998, United Kingdom and Austria; 1999, Germany and Finland; 2000, Portugal and France.

The schedule of Council meetings is agreed upon for each Presidency, and their frequency depends upon the issues; the General Affairs, Economic and Finance, and Agriculture Councils meet every month, whereas more specialized Councils such as the Environment meet two or three times a year. Each Presidency determines priorities, sets agendas for the next six months, and elects the necessary working groups. At the end of each Presidency, heads of states and governments meet in a European Council. The European Council meeting is not a Community institution as such, but is a "summit" that provides the driving force behind the Union.

The main role of the Council is to adopt legislation that has been proposed and amended through the decision-making procedures of the institutions of the European Union. Depending on the legal basis of the proposal, the vote to adopt legislation must be either unanimous, a qualified majority, or a simple majority.

The European Parliament

The European Parliament consists of 626 members elected by universal suffrage in each Member State for a five-year period. The role of the Parliament differs from that of a national parliament, owing to the European Union's institutional system. It originally had only a consultative role, but its powers have been gradually increased, and it now holds colegislative status. It also exercises important democratic control over the execution of Community law and further contributes to transparency and openness through its public plenary sessions.

Main Legal Instruments

The Council and the Commission may make regulations, issue directives, make decisions, make recommendations, or deliver opinions. This discussion focuses on the central types of instruments used: regulations and directives.

Regulations

A Community regulation has general application, is binding in its entirety, and is directly applicable in all Member States. It is used when there is no need to allow some flexibility at Member State level, or where flexibility is unacceptable. It is extensively used in the area of agriculture and market regulations, and is increasingly used in the environmental field, especially when the Commission needs to collect and centralize data (e.g., the Council Regulation 793/93 of March 23, 1993 on the Evaluation and Control of the Risks of Existing Substances). Because a Community regulation is already binding, there is no need for Member States to transpose it into their national laws.

Directives

Directives are addressed to Member States and are binding only as to the result to be achieved. A directive must be transposed into the Member States' national laws by a certain deadline. However, each authority may choose the form and the method to achieve the result within its own constitutional and legislative framework. Therefore, some discretion is left to the Member States as to the implementation of directives.

Adopting Regulations and Directives

The Commission, the Council, and the Council and the Parliament together, can adopt regulations and directives. The Commission's instruments implement preexisting Council or Council and Parliament regulations and directives. They set out the procedures that the Commission must follow to adopt measures, which are Commission Regulations or Commission Directives. Technical Adaptations to the Dangerous Substances Directive are made in this way.

For Council and/or Parliament measures, the aim and the content of the measures determine their legal basis, which in turn determines the decision-making procedure used to adopt them. These measures are adopted by the Council following consultation with the European Parliament, in cooperation with it or by a codecision. The difference between the procedures involved lies in the influence of the European Parliament in the formation of a measure and in the voting majority required of the Council of Ministers in adopting it.

Disputes between the Institutions arise in selecting the legal basis, as it is related to the aim and the content of the measures. These disputes may be brought to the European Court of Justice.

THE DANGEROUS SUBSTANCES DIRECTIVE

History

The Dangerous Substances Directive marked the entry of the European Community into product health and environmental legislation, and it can be seen as a landmark in its field.

The Dangerous Substances Directive (67/548/EEC) was adopted in 1967 (European Council, 1967). Since then, it has been modified 30 times with 8 amendments and 22 adaptations to technical progress. Major modifications include the 1979 Sixth Amendment, which consolidated the classification, packaging, and labeling provisions and also introduced a notification scheme for new substances, including new polymers (European Council, 1979). The 1992 Seventh Amendment (European Council, 1992) addressed the same issues as the Sixth Amendment (i.e., classification and labeling and notification). It also brought substantial changes on research and development exemptions, introduced the concept of risk assessment in the notification process of new substances, and included for the first time the definition of a polymer.

Impact on Polymers

The Dangerous Substances Directive (DSD) can be characterized as a classification, packaging, and labeling directive, a notification of new substances directive, and a cataloguing substances directive (inventory). The status and the impact of DSD on polymers is reviewed in this section.

Polymers and the European Inventory of Existing Chemical Substances (EINECS)

Unlike the U.S. TSCA inventory, EINECS is a closed inventory, listing only commercial substances on the European Community market between 1971 and 1981. EINECS excluded the reporting of polymers, polyadducts, polycondensates (only monomers as starting materials were reportable), and post-reacted polymers. Post-reacted polymers were defined as polymers that have been post-reacted (i.e., have undergone reactions after polymerization). The monomeric substances from which the polymer was manufactured and the post-treating reagent(s) could be reported separately. However, the basic documents on constructing EINECS, which provided the criteria for reporting, and the Sixth Amendment did not contain any definition of a polymer.

For about ten years, conflicting views existed among industry representatives, governmental authorities, and the Commission on definitions of
polymers and oligomers, what to do with post-reacted polymers, and
whether definitions should be provided at all. The definition of a polymer
in the Seventh Amendment and the setting up of the so-called No Longer
Polymers (NLP) list has helped to put an end to this uncertain situation
(see below).

Classification and Labeling of Polymers

Under current law, polymers, like any other substances, may need to be
classified and labeled if they are considered hazardous for health or the
environment according to criteria defined in DSD. The classification for
environmental affects is currently restricted to aquatic hazards.

Environmental classification and labeling was first introduced in 1991
and was modified in 1993. The date of effect in the Member States was July
1, 1994. Though provision was made for classification of hazardous properties associated with other environmental compartments, with the exception
of the ozone layer no criteria beyond aquatic hazards have yet been given.

Relatively simple environmental classification criteria with distinct
cutoff points are used for classification of substances for aquatic hazard. The
classification criteria are based on inherent toxicity of substances to fish,
Daphnia, or algae and also the potential to degrade or bioaccumulate.
Degradation is primarily defined in terms of ready biodegradability and
follows test cutoffs and criteria. Data on abiotic degradation or the lack of
chronic toxicity may also be considered if the substance is not readily
biodegradable. Bioaccumulation is defined in terms of the octanol/water
coefficient ($\log P_{ow}$), but actual measured bioconcentration values may be
used if available. Toxicity in aquatic species is defined in terms of acute
LD_{50} or EC_{50} values. The following criteria are used to warn of short-term
and long-term dangerous properties:

- If acute toxicity (LC_{50} or EC_{50}) is less than 1 mg/L, the substance is
  assigned a toxicity "risk" phrase R.50 (Very toxic to aquatic organisms) and the "N" symbol of danger (i.e., label with dead fish/dead
  tree) regardless of its degradability. It may be assigned an R.53 phrase
  (May cause long-term adverse effects in the aquatic environment) if it
  is deemed to be a long-term danger to the aquatic environment.
- For substances with acute toxicity between 1 mg/L and 10 mg/L, R.51
  plus R.53 (Toxic to aquatic organisms. May cause long-term adverse
  effects in the aquatic environment) is used. The "N" symbol of danger
  is assigned if a substance is deemed to be a long-term danger to the

aquatic environment. Lack of potential for long-term danger is typically indicated with positive ready biodegradability data (e.g., >60% oxygen depletion or carbon dioxide generation in 28 days) and log P_{ow} (<3.0), but other types of data may be considered.

- For acute toxicity between 10 mg/L and 100 mg/L, R.52 plus R.53 (Harmful to aquatic organisms. May cause long-term adverse effects in the aquatic environment) is assigned, if the material is not readily biodegradable. Otherwise, no label is required.
- As exceptions, some substances may carry single risk phrases (R.52 or R.53). Although rules for application of these single risk phrases are not fully defined, poorly water-soluble substances may require R.53. In any case, substances classified R.52 or R.53 will not be required to carry the "N" symbol and "dangerous for the environment."

Criteria applicable to the aquatic environment currently apply only to substances, but will be extended to preparations (i.e., blends) within the next two or three years upon adoption and implementation of the first amendment to the Dangerous Preparations Directive (DPD).

Premarketing Notification (PMN) of Polymers

Under the U.S. Toxic Substances Control Act (TSCA), all new polymers must be notified, but some exemptions are foreseen. The TSCA Inventory also lists polymers. In the E.U., polymers containing in combined form less than 2% of any substance not listed on EINECS do not require notification. Conversely, if a polymer contains more than 2% of a new substance in a combined form, it is a new polymer that must be notified and is subject to an assessment of its risks to humans and the environment, like all other new notified substances. Therefore, the key to the issue is the definition of "polymer."

In January 1990, an Organisation for Economic Co-operation and Development (OECD) Group of Experts met in Toronto and agreed on a definition of "polymer" that could be used to differentiate polymers from discrete substances in the context of notification. This definition was known for some years as the "Toronto definition." The following definition was adopted in the Seventh Amendment:

Polymer means a substance consisting of molecules characterized by the sequence of one or more types of monomer units and comprising a simple weight majority of molecules containing at least three monomer units which are covalently bound to at least one other monomer unit or other reactant and consist of less than a simple weight majority of molecules of the same molecular weight. Such molecules must be distributed over a range of molecular weights wherein differences in the molecular weight are primarily attributable to differences in the number of monomer units. In the context of this definition a "monomer unit" means the reacted form of a monomer in a polymer.

The Seventh Amendment contains Annex VIID, which identifies polymers for which the full notification (i.e., "base set") is necessary. As outlined in detail in the Annex, the base set can include:

- Identity of the manufacture and identity of the notifier.
- Identity of the substance (e.g., chemical name).
- Information on the substance (e.g., production process and exposure data).
- Physical-chemical properties (e.g., density).
- Toxicological studies (e.g., irritation data).
- Ecotoxicological studies (e.g., acute toxicity in aquatic species).
- Possibility of rendering the substance harmless (e.g., recycling potential).

The Annex also identifies polymers of "low concern" on the basis of three criteria for which a reduced test package is acceptable:

For non-readily degradable polymers placed on the Community market ..., the following criteria define those polymers for which a reduced test package is acceptable:

(a) High number-average molecular weight (MW_n); authorities receiving the notification decide on their own responsibility whether or not a polymer satisfies this criterion,
(b) Extractivity in water: $< 10\,mg/L$ excluding any contribution from additives and impurities,
(c) Less than 1% with $MW < 1000$; the percentage refers only to molecules (components) directly derived from and including monomer(s), excluding other components (e.g., additives or impurities).

Mammalian toxicity and ecotoxicity tests are required for polymers on a case-by-case basis. For ecotoxicity tests, the authorities receiving the notification may require certain tests to be carried out based on the presence of reactive groups, structural/physical characteristics, knowledge concerning the properties of low molecular weight components of the polymer, or exposure potential. They may, in particular, require light-stability data (i.e., if the polymer is not specifically light-stabilized) or long-term extractivity information (i.e., leachate test). Depending on the results of this latter test, any toxicity test on the leachate may be requested on a case-by-case basis.

Waste Issues

European waste law has been at the center of major changes for several years. The following discussion provides a brief background on where the Law stands and focuses on the recently adopted Packaging and Packaging Waste Directive.

European Waste Legislation

Waste law has undergone major changes over the past few years, which now need to be reflected at the national level. These changes, among others, can be found in a Council Resolution in 1990 that endorsed the Commission's Communication on Waste Management Strategy. Its purpose was to set out principles, policy objectives, and actions that the Commission intended to follow in developing proposals for legislation and other activities in subsequent years.

Definitions of wastes, disposal, and recovery have recently been modified. The criteria of proximity and self-sufficiency have been adopted and recycling and reuse promoted, all with the aim of ensuring a high level of protection of the environment. Written in a descriptive manner, European waste law currently consists of about 30 legal instruments in force and a few others in preparation. Rules concern the movement of wastes, the incineration of municipal wastes, the incineration of hazardous wastes, landfilling, and incineration at sea. There are specific directives for hazardous waste, polychlorobiphenyl and polychloroterphenyl wastes, waste oils, sewage sludge, beverage containers, waste from the titanium dioxide industry, and packaging.

Packaging Waste

Polymers, including plastics, will be affected when the directive on packaging and packaging waste takes effect (European Parliament and Council,

1994). Packaging and packaging waste is a fairly new area of legislative activity in the E.U., but many initiatives have been taken by individual countries (e.g., Germany, The Netherlands, Austria, Sweden, and France) to reduce the amount of packaging waste. This represents an effort to minimize the generation of packaging waste and to increase the recovery of waste that is produced, thus reducing quantities in landfills.

The directive sets global target ranges for recovery and recycling of packaging waste, and minimum recycling targets for packaging material. It covers all packaging types (i.e., household, commercial, and industrial). On a by-weight basis, 50% to 65% of packaging waste shall be recovered from waste streams no later than five years after the implementation of the directive (i.e., by the year 2001). Some 25% to 45% of the packaging waste is to be recycled, with the rest incinerated by energy recovery technologies or otherwise recovered by alternative methods. In addition, there is a 15% minimum recycling rate for each material. Standards and markings for packaging must be developed within two years of implementation.

For funding recovery operations, the directive allows the use of economic and fiscal instruments. However, in terms of impact, the industry will have to develop the capacity to recycle a minimum of 2 million metric tons of packaging plastics by the year 2001 without counting the waste of other applications (e.g., automotive, electric, and electronic). Current energy recovery methods are sufficient but are not accepted. Recycling capacities are not sufficient. Environmental pressures and the cost of recovery may affect the growth of the plastics industry by minimizing packaging and material substitution.

REFERENCES

European Council. 1967. Council Directive 67/548/EEC of 27 June 1967 on the approximation of laws, regulations, and administrative provisions relating to the classification, packaging and labelling of dangerous substances. *Official Journal of the European Communities*, OJ No. 196, 16 August 1967.
European Council. 1979. Council Directive 79/831/EEC of 18 September 1979 amending for the sixth time Directive 67/548/EEC on the approximation of laws, regulations, and administrative provisions relating to the classification, packaging and labelling of dangerous substances. *Official Journal of the European Communities*, OJ No. L 259, 15 October 1979.

European Council. 1992. Council Directive 92/32/EEC amending for the seventh time Directive 67/548/EEC on the approximation of laws, regulations, and administrative provisions relating to the classification, packaging and labelling of dangerous substances. *Official Journal of the European Communities*, OJ No. L 154, 5 June 1992.

European Parliament and Council. 1994. European Parliament and Council Directive 94/62/EEC of 20 December 1994 on packaging and packaging waste. *Official Journal of the European Communities*, OJ No. L 365, 31 December 1994.

PART *4* | *New Directions*

Chapter 15

Environmentally Biodegradable Polymers

■ GRAHAM SWIFT
Rohm and Haas Company

INTRODUCTION

Environmental opportunities in the commercial polymer market have prompted manufacturers to search for suitable biodegradable polymer products. Plastics are polymers that are generally designed to combine light weight with excellent resistance to mechanical, biological, and chemical resistance. Typical plastics show virtually no degradation even after years of exposure in natural environments and landfills. Polycarboxylates are water-soluble acrylic polymers, some of which were developed for detergent formulations to replace phosphates. These polymers are of very low ecotoxicity (Chapter 6; Hamilton et al., 1996), but are not generally highly biodegradable. The challenge is to identify opportunities where plastics, polycarboxylates, and other useful polymers can be made environmentally biodegradable while remaining both commercially competitive and compatible with existing waste management programs.

The success of a biodegradable polymer development program requires:

- A reasonable understanding of the fate of the new polymer in the environment.
- Chemical performance properties similar to, or better than, those of the polymers that the biodegradable polymers are designed to replace.
- Reasonable cost.

DEFINITIONS

It is important to recognize the need for accurate definitions and for test methods consistent with those definitions. Much of the earlier research and development of environmentally biodegradable polymers was confusing to legislators and the public because the researchers had no clearly defined goals, no definition of terms, and little understanding of the several degradation pathways operating in the environment. Often biodegradation was assumed to include all degradation pathways found in the natural environment and/or physical-chemical effects such as surface erosion. No standardized test methods were available. However, this early work helped establish guidelines for correlating polymer structure with specific environmental degradation pathways.

International cooperation on definitions and protocols is rapidly emerging through the efforts of such organizations as the Japanese Biopolymer Society, the American Society for Testing and Materials (ASTM), the International Organization of Standards (ISO), the Institute for Scientific Research, the Bio/Environmentally Degradable Polymer Society, the Compost Council in the United States, the National Corn Growers Association, and the Organisation for Economic Co-operation and Development (OECD). There are several academic institutions and government research centers devoted to environmentally degradable polymer research, particularly biodegradable polymer research. For example, the cooperative research programs and industrial activities in Japan were recently reviewed (Lenz, 1995) and suggested a strong interest in biodegradable polymers. The U.S. National Science Foundation provided a delegation to Japan in 1993 for joint discussion of biodegradable polymer chemistry.

Environmental degradation includes biodegradation, photodegradation, oxidation, and/or hydrolysis leading to some change in polymer structure and physical properties. What is not clearly understood is the degree of degradation that it is reasonable to expect a polymer to undergo for satisfactory disposal in the environment. One of the first steps toward international consensus on terminology was taken by ASTM (1993), with these definitions:

- *Degradable polymer:* a polymer designed to undergo a significant change in its chemical structure under specific environmental conditions, resulting in a loss of some properties that may vary as measured by standard test methods appropriate to the polymer and the application in a period of time that determines its classification.
- *Biodegradable polymer:* a degradable polymer in which the degradation results from the action of naturally occurring microorganisms such as bacteria, fungi, and algae.

- *Hydrolytically degradable polymer:* a degradable polymer in which the degradation results from hydrolysis.
- *Oxidatively degradable polymer:* a degradable polymer in which the degradation results from oxidation.
- *Photodegradable polymer:* a degradable polymer in which the degradation results from the action of natural light.

The ASTM definitions only help differentiate degradation pathways; they give no indication of how degradation pathways impact the overall potential of a polymer to cause ecological effects. The interrelation of degradation pathways is shown schematically in Figure 15.1.

Biodegradation can play the key role in ultimately defining environmentally acceptable polymers. In Figure 15.1, all degradation pathways lead to fragmentation of the polymer backbone chain. However, the polymer fragments and residues (i.e., metabolites) may remain in the environment, or they may completely biodegrade. Rapid biodegradation is a strong indication of acceptability for disposal in a suitable environment.

The concentration of polymers and residues in specific environment compartments (e.g., soil, water, sediments, etc.) can be estimated by measuring the rate(s) of biodegradation and the identity of the residues. The estimated residue concentrations can be compared with threshold concentrations for toxicity in the same environment. An environmentally biodegradable polymer degrades in the environment by any specified pathway, whereas an environmentally acceptable biodegradable polymer leaves no ecologically harmful fragments and residues.

The degree of polymer biodegradation can be measured according to the mass of carbon dioxide and/or methane evolved, degradation products (e.g., monomers) released, and polymer carbon converted into biomass.

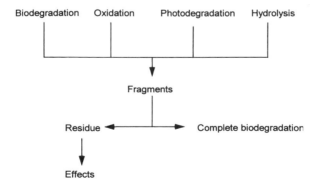

Figure 15.1.   Environmental degradation of polymers.

Chapters 3 through 5 provide details on measurements of polymer biodegradability.

SIGNIFICANCE

Using environmentally biodegradable polymers is only one of many options in the waste management of polymers. Biodegradable polymer programs often compete with incineration, recycling, and landfill management. Environmentally biodegradable plastics are likely to be favored for situations where the recovery of conventional plastics for recycling or incineration is difficult and/or costly. However, the recovery of water-soluble polymers from wastewater is difficult; so environmental biodegradability represents a more realistic opportunity for such polymers than does recovery.

There are important differences in waste management options for polymers across international markets. Japan continues to favor the incineration of plastics; there incineration is viewed as part of energy recycling for petroleum feedstocks. Little plastics recycling is practiced in Japan at present (Lenz, 1995), but the trend for recycling use is upward. In contrast, European regions tend to favor recycling and composting of some plastics. In the United States, landfill use is prevalent, but recycling of suitable waste plastics is rapidly gaining favor. Composting of biodegradable polymers also is being considered in the United States.

WASTE MANAGEMENT

Development of environmentally acceptable biodegradable polymers is a primary goal for synthetic chemists. Achievement of this goal requires suitable test protocols to simulate expected disposal methods.

Biodegradation is currently the most practical degradation pathway for water-soluble polymers. Although there are opportunities for degradation of some water-soluble polymers by photodegradation, oxidation, or hydrolysis, these pathways are not generally practical for water-soluble polymers because each requires specific activation methods. On entering a wastewater treatment facility, a water-soluble polymer may remain in solution, sorb completely to the suspended solids and/or sludge, or partition within solution and sorbed phases. The degree of partitioning depends on polymer characteristics such as structure, functionality, water solubility, and molecular weight. Polymers that remain in solution may either biodegrade completely or pass through within hours unchanged. Other polymers, which sorb strongly to suspended solids, are retained in the wastewater treatment facility for a considerably longer period of time, usually one to two weeks. If not completely biodegraded in that time, the sorbed fractions eventually are removed with sludge and transferred to landfills, land application sites for soil amendment, compost, or incinerators. An assessment of polymer fate and effects in natural environments and disposal sites (e.g., landfills) may be required. Therefore, complete biodegradation during wastewater treatment, as offered by some new water-soluble poly(amino acids) (see below), may become a preferred means of disposal of waste-soluble polymers.

Plastics are often recoverable as solids after use. Disposal routes include landfilling, recycling, incineration, and composting with suitable organic waste. Landfilling and composting have provided opportunities for development of environmentally biodegradable polymers. Many landfill sites are relatively inactive repositories for plastics (Hamilton et al., 1995); however, there are opportunities for biodegradable polymers in anaerobically active landfills that are managed for gas-energy production. Composting could represent a major opportunity for plastics such as those used in fast-food packaging and utensils. Plastics for agricultural films and personal hygiene products represent other opportunities for the development of biodegradable polymers.

TEST METHODS

Test methods for environmentally biodegradable polymers should be designed to help predict ecological toxicity or fate by simulating real-world (i.e., natural environment or disposal site) conditions during use or after disposal. However, it is often desirable to use accelerated conditions to rapidly establish the relative potential for degradability or to use minimal conditions to positively identify only polymers with high degradation potential. ASTM has led in programs to develop standard methods for testing environmentally biodegradable polymers.

Photodegradation and Oxidation

Photodegradation and oxidative degradation of polymers are measured by combining an ASTM test practice for exposure, such as exposure to light, with an ASTM test method for measuring a property change, such as tensile strength (ASTM D882-83), tear strength loss (ASTM D1922-67), or impact resistance (ASTM D1709-85). Standard test practices recently developed by ASTM for photodegradable and oxidatively degradable polymers (ASTM, 1993) are listed in Table 15.1.

Measurements of weight loss and physical-chemical changes have value where a certain market requires that biodegradation promote fragmentation. For example, in compost films and fishing gear, physical-chemical changes from biodegradation may be followed by specific tests for tear strength, impact resistance, and tensile strength.

Biodegradation

Factors in Biodegradation Test Design and Selection

Aerobic biodegradation is a relatively common mechanism of wastewater treatment, and media from wastewater treatment processes have received the most attention in test protocol development. Test media can also include compost, water, or soil. Anaerobic test media environments are available for water-soluble polymers and plastics that may reach anaerobic landfills or septic treatment systems. The choice of test media should be governed by the anticipated disposal method, the use environment, and whether accelerated tests are desired.

Microbial populations vary widely according to location and season. It is occasionally advisable to run biodegradation tests according to multiple sampling locations and seasons. Sometimes microbial acclimation to a polymer substrate is necessary before significant biodegradation occurs. Temperature and pH may influence biodegradation of a particular polymer, especially if polymer hydrolysis or oxidation is necessary to facilitate biodegradation.

Table 15.1. ASTM standard practices for photodegradation and oxidation

Test Number	Standard Practice
D3826-91	Degradation End Points: A Tensile Test
D5071-91	Xenon Arc Exposure
D5208-91	Operation of a Fluorescent UV and Condensation Apparatus
D5272-92	Outdoor Exposure of Photodegradable Plastics
D5437-93	Marine Weathering, Floating
D5510-94	Heat-Aging of Oxidative Degradable Plastics

Polymer test concentrations occasionally may have a strong influence on biodegradation results. High test concentrations may inhibit respiration of the test medium, and low concentrations may not permit analytical detection of biodegradation. Before the biodegradation test begins, it is wise to establish a suitable test concentration with a range-finding method to measure the potential for respiration inhibition. As an alternative, polymer radiolabeling often allows the use of low polymer concentrations to permit highly sensitive analytical detection of biodegradation without inhibition of the test medium.

Test protocols for extractable polymer residues and fragments are not yet widely developed. However, Scholtz (1991) published a paper suggesting an improved evaluation of residues from water-soluble polymers, and ASTM developed a protocol (ASTM D5152-91) for isolating water-soluble extractable residues and transferring them to standard toxicological tests.

Biodegradation testing with commercial polymers has greatly improved over the past few years. However, the available test methods still should be regarded as screening tests just to suggest the degree of potential for biodegradation. Correlation with actual environmental conditions is essential, and the question of degradation of polymer fragments and residues must be carefully addressed.

Burial and Microbial Growth Tests

Much of the early work in polymer biodegradation involved burial and microbial growth tests to indicate susceptibility of plastics to biodegradation. Microbial growth tests measure microbial bloom with a polymer in agar dispersion, as an indicator of biodegradation. Such tests still are used to screen for susceptibility to biodegradation with plastics and water-soluble polymers. However, microbial growth tests are susceptible to false positive results caused by biodegradation of plastic additives, plasticizers, and mold contamination.

Microbial growth tests are described by the ASTM Standard Methods G21-70 and G210-76. Both methods were originally intended to measure

Table 15.2. Biodegradation ratings in fungal and bacterial growth tests

Rating	Growth
0	No visible growth
1	< 10% of surface with growth
2	10–30% of surface with growth
3	30–60% of surface with growth
4	60–100% of surface with growth

whole product resistance to microorganisms. ASTM G21-70 uses fungal microorganisms such as *Aspergillus niger, Aspergillus flavus, Chaetomium globosum,* and *Penicillium funiculosum.* Standard Method G22-76 uses bacteria such as *Pseudomonas aeruginosa.* It is permissible to substitute test species in these tests. Microbial growth density is rated 0 to 4. The test results are interpreted as shown in Table 15.2, with higher ranking of microbial growth correlating with higher assumed susceptibility to biodegradation.

Clendinning and others have demonstrated the potential biodegradability of polycaprolactone by burial in soil and, following its disappearance, visual inspection over time (Clendinning et al., 1974). The burial method is still used as a "field test" for natural environments (Sawada, 1994).

Quantitative Tests

A need for quantitative biodegradation tests protocols became evident after the development of starch–polyolefin blend polymers, based on work by Otey et al. (1987). These polymers were suggested to be eventually "completely" biodegradable because the polyolefin film product disintegrated in soil. However, polyolefins (e.g., polyethylene) now are known to be virtually nonbiodegradable; only the starch in the film product was susceptible to biodegradation.

Many of the new test methods being developed for plastics originate from aquatic fate testing by the detergent industry. Swisher (1987) has summarized tests from organizations such as the U.S. Environment Protection Agency (U.S. EPA) and the OECD. Major tests for biodegradation of polymers include the following:

Table 15.3. Degradation testing protocols

Test number	Medium	Measurement
ASTM D5209-92	Aerobic sewage sludge	Carbon dioxide
ASTM D5210-92	Anaerobic sewage sludge	Carbon dioxide, methane
ASTM D5247-92	Aerobic microorganisms	Molecular weight
ASTM D5271-93	Aerobic activated sludge	Oxygen/carbon dioxide
ASTM D5338-92	Controlled composting	Carbon dioxide
ASTM D5437-93	Marine floating test	Physical properties
ASTM D5509-94	Simulated compost	Physical properties
ASTM D5511-94	Anaerobic biodegradation	Carbon dioxide/methane
ASTM D5412-94	Simulated compost	Physical properties
ASTM D5525-94A	Simulated landfill	Physical properties
ASTM D5526-94	Accelerated landfill	Carbon dioxide/methane

- Measurement of chemical changes and residue products formed.
- Test environments that represent conditions from natural, simulated, accelerated, disposal or proposed use scenarios.
- The effects of plastic form and shape as well as the concentration of polymer residue fragments on biodegradation.

As shown in Table 15.3, ASTM (1993) has provided test protocols for assessing the biodegradability of polymers. Biochemical oxygen demand (BOD) compared to maximum theoretical uptake of oxygen (ThOD) for complete oxidation is often used to screen for rapid aerobic biodegradation potential. Measurement of carbon dioxide release also is a screening method for aerobic biodegradation potential. The presence of readily oxidizable chemicals can lead to false positive results for biodegradability in BOD tests, as can polymer impurities or biodegradable additives in carbon dioxide release tests. False negative results are also possible with the wrong choice of inoculum for biodegradation testing. For example, the inherently low biodegradation activity in BOD tests also can lead to false negative results. Detailed information on important biodegradability tests with polymers is provided in Chapters 3 through 5.

For completely water-soluble (i.e., solubility without additive) polymers (e.g., polycarboxylates), dissolved organic carbon (DOC) can be used to measure polymer mass biodegraded as an alternative to measuring BOD, carbon dioxide release, or biomass produced. For plastics, carbon dioxide release is the most common measure of biodegradation. For total accountability, radiolabeled polymers can be used to obtain the mass balance for the polymer biodegradation process. This method may be the only option when the polymer is formulated with other biodegradable components, is of low water solubility, and/or will cause toxicity in the test medium.

It should be noted that BOD, DOC, and related screening-level tests should not be used to suggest a lack of biodegradation potential of polymers under realistic test conditions. Further, there are numerous media-specific biodegradation tests for plastics, including compost (see Chapter 5) and anaerobic bioreactor testing (Gu et al., 1992).

DEVELOPMENT

General Aspects

The hydraulic residence time for polymers that will not sorb to wastewater treatment sludge may be only a few hours. However, polymers that sorb to wastewater treatment sludge may have a residence time for degradation in the range of 1 to 2 weeks. Plastics for disposal in compost have about 4 weeks for complete biodegradation. As litter, plastics should degrade in a reasonable and measurable time frame. Developments in producing environmentally acceptable biodegradable polymers are outlined below.

Photodegradable and Oxidative Degradable Polymers

Photosensitive groups such as carbonyl groups may be introduced into a polymer backbone or a side-chain directly by the copolymerization of carbon monoxide or a vinyl ketone, respectively, with olefinic and other unsaturated monomers (Guillet and Americk, 1971; Hartley and Guillet, 1968; Heskins and Guillet, 1968). Incorporation of external photosensitizers and pro-oxidants such as metal salts (Griffin, 1977) facilitates photodegradation. Patented chemical additives based on benzophenone, ketones, ethers, mercaptans, or polyunsaturated fats also aid polymer photodegradation. Common applications for photodegradable polymers include polyethylene–carbon monoxide polymers that are used in beverage holders and agricultural films.

Recently, photodegradable polymer chemistry has been coupled with biodegradable aliphatic polyester chemistry. Based on the work of Bailey and others with ketene acetals (Bailey et al., 1979) Exxon developed a patented process to polymerize ethylene, carbon monoxide (CO), and 2-methylene-1,3-dioxapane to introduce an ester functionality, as shown in Figure 15.2, by ring-opening rearrangement of the ketene acetal monomer. Another manufacturer, Quantum, has used Baeyer-Villigar oxidation to convert a conventional polyolefin–carbon monoxide polymer to a potentially photodegradable polyester, as shown in Figure 15.3.

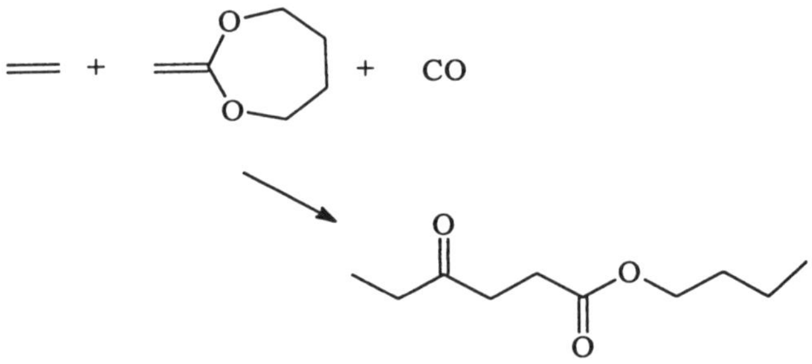

Figure 15.2. Ring-opening rearrangement of a ketene acetal to form an ester.

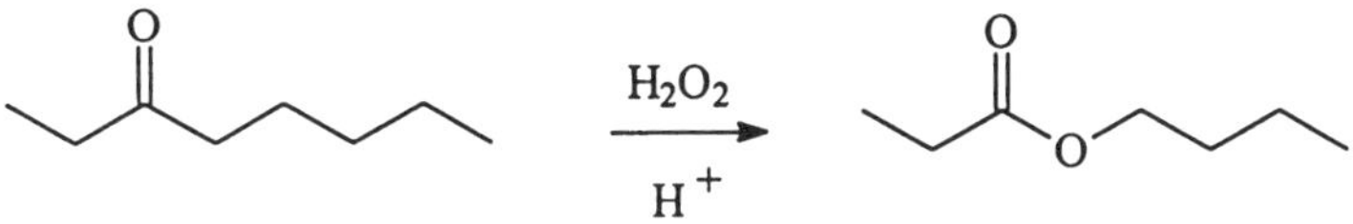

Figure 15.3.   Conversion of a polyolefin–carbon monoxide polymer to a potentially photo/biodegradable polyester.

Biodegradable Polymers

Broad conclusions based on design experience with biodegradable synthetic and natural polymers are as follows:

1. Natural polymers are considered biodegradable.
2. Chemically modified natural polymers may be biodegradable, depending on the extent of modification.
3. Synthetic addition polymers with carbon chain backbones usually do not biodegrade at molecular weights greater than about 500.
4. Synthetic addition polymers with heteroatom backbones (e.g., oxygen and nitrogen) may be biodegradable, depending on other structural features.
5. Synthetic step growth, or condensation, polymers are generally biodegradable to some extent, with degradation depending on:
   (a) Chain coupling: ester > ether > amide > urethane.
   (b) Molecular weight: lower MW polymer degradation is generally faster than higher MW polymer degradation.
   (c) Morphology: amorphous polymer degradation is faster than crystalline polymer degradation.
   (d) Hydrophilicity: aliphatics are more biodegradable than aromatics.
6. Water solubility does not guarantee biodegradability.
7. High carboxyl functionality may impair biodegradability.
8. Virtually all synthetic polymers that are likely to biodegrade are analogous in structure to one or more natural polymers.

Some examples of synthetic polymers considered to be biodegradable include polyethers (Kawai, 1987), poly(lactic acid) (Keeler, 1991), and polycaprolactone (Clendinning et al., 1974).

Aliphatic forms of polyesters have been found to have significant potential for biodegradation, a finding consistent with recent patent developments by Showa High Polymers and Eastman. Future development of biodegradable polymers probably will be dominated by synthetic chemicals that produce polymers with structures known to promote biodegradation, as well as modified natural polymers and blends and polymers from natural sources.

Synthetic Polymers

Bailey's work (e.g., Bailey et al., 1979) with ketene acetals (see above) and Monsanto's patented work on polyacetals and polyketals are highly informative. This research helped inspire the development of photobiodegradable polymers with polyester linkages and American Cyanamid's biodegradable copolymers of acrylic acid with ketene acetals. Chemists at Monsanto used a pH trigger to develop biodegradable polyacetal and polyketal carboxylates as water-soluble replacements for phosphates in detergents. Polyacetals and ketals are stable under the basic pH conditions where they are used in detergents. However, after dilution lowers the pH to about 7, the same polymers will rapidly hydrolyze to a monomer that is rapidly biodegradable. The current weakness of these approaches is the cost of raw materials.

Patented aliphatic polyesters such as caprolactone from Union Carbide, succinate-polyols from Showa High Polymers and Eastman, and poly(lactic acid) from Cargill are biodegradable plastics. A low molecular weight poly(aspartic acid) that was patented by the Rohm and Haas Company is highly biodegradable when precise control is exercised during polymerization to maintain linearity of the polyaspartic chain. The poly(aspartic acid) represents the first polymer that can completely biodegrade within the residence time of aerobic wastewater treatment. The polymer sorbs to aerobic wastewater treatment sludge, which typically has a residence time of about 14 days. Complete biodegradation of poly(aspartic acid) was observed within approximately 10 days. The success of this polymer has encouraged much research activity with poly(amino acids). Many important patents have appeared worldwide for polyamino acids as dispersants in detergents, superabsorbents, and other applications.

Modified Natural Polymers and Blends

Starch is a low cost, readily available raw material that has been widely evaluated as a raw material for biodegradable water-soluble polymers and plastics. There is considerable interest in the synthesis of starch-based detergent polymers, including polyglycosides and carboxylated polymers (e.g., Matsumura et al., 1993). However, significant cost issues currently exist in producing carboxylated starch polymers. The carboxylated form of starch is produced by oxidation, which is typically done at low solids content and with expensive reagents such as periodate and perchlorate. However, catalytic processes based on metal catalysts or redox reactions are emerging.

Plastics based on starch blends continue to be evaluated, especially those based on poly(ethylene oxide) and poly(vinyl alcohol). Significant patents have been awarded to manufactures such as Novamont, Novon, and National Starch. Synthesis of other modified natural polymers includes work with modified lignin (Meister et al., 1991), cellulose acetates (Gu et al., 1992), and other cellulosics such as the propionates, hydroxyethyl cellulose, and the carboxymethylated derivatives.

Despite all the activity in synthetic polymer research, there is not enough evidence that many modified natural polymers are completely biodegradable during biological waste treatment and/or in the natural environment. Much remains to be done in this very promising area, including additional evaluation with standard biodegradation test methods.

Natural Polymers

A major advantage of biodegradable natural polymers is that they are based on renewable resources. Some natural polymers are already available in specialty applications; for example, xanthan gum is a commercial thickener. Other potentially commercially viable natural polymers are polyhydroxyalkanoates, bacterial celluloses, alginates, hyaluronic acid, and spider silk.

There is a wide range of polyhydroxyalkanoate structures. Certain bacteria produce these polyesters as storage materials during periods of nutrient stress. Significant research has involved polyhydroxyalkanoates (Doi et al., 1990). These bacterial polyesters (see Figure 15.4) are among exciting developments in polymer science in recent years.

Natural polymers are expensive, and the question of their commercial viability has not been answered. It is significant to note that the time from discovery to the relatively small-scale production of a polyhydroxyalkanoate product was 60 years.

Physical properties of a polyhydroxyalkanoate sold under the name Biopol are similar to those of polypropylene. However, the current cost for Biopol is considerably higher than that of polypropylene. Biopol is

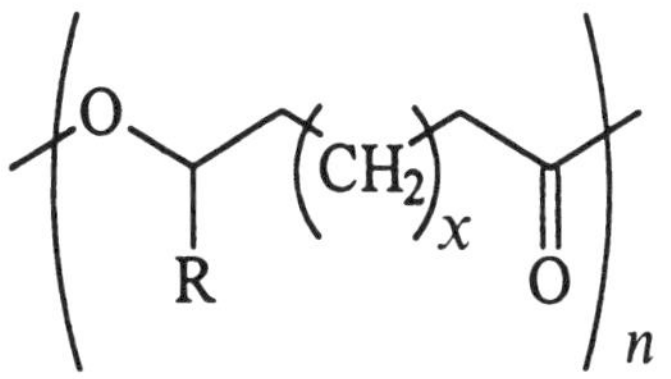

Figure 15.4.   Polyhydroxyalkanoate structure.

produced by fermentation, a process that is a relatively new approach for commercial polymer production. Higher production volume eventually may bring down per-unit fermentation costs. Also, there are innovative efforts to transfer bacterial polyester production genes to plants for enhanced polyester yield. Genetic engineering may markedly decrease the cost of natural polymers.

PRODUCT EXAMPLES

The efforts described above have led to polymer products that completely or partially degrade under environmental conditions. Some of these products are listed in Table 15.4. In particular, water-soluble poly(amino acids) such as poly(aspartic acid) show excellent promise as environmentally acceptable biodegradable polymers.

Table 15.4.   Polymer products with degradation potential

Polymer	Developer[a]	Degradation
Cellophane	Flexel	Biodegradation
Cellulosics	Rhone-Poulenc Eastman	Biodegradation
Poly(aspartic acid)	Rohm and Haas	Biodegradation
Polycaprolactone	Union Carbide, Solvay	Biodegradation
Polyesters	Showa High Polymer, Zeneca	Biodegradation
Polyethylene-CO	Dow	Photodegradation
Polyethylene glycol	Union Carbide, Dow	Biodegradation
Poly(lactic acid)	Cargill, Ecochem, Biopak, Mitsui Toatsu, Shimadzu	Hydrolysis, biodegradation
Polyolefin-activator	Plastigone	Photodegradation
Poly(vinyl alcohol)	Rhone-Poulenc, Air Products, Kuraray, Hoechst	Biodegradation
Starch-based	Novamont	Biodegradation
Starch-activator	Ecostar	Photodegradation, biodegradation
Starch foam	National Starch	Biodegradation

[a]Examples only; not inclusive list.

References

ASTM. 1993. *ASTM Standards on Environmentally Degradable Plastics.* ASTM Publ. Code No. 03-420093-19. Philadelphia, PA: American Society for Testing and Materials.

Bailey, W. J., P. Y. Chen, Y. Paul, W.-B. Chiao, T. Endo, L. Sidney, N. Yamamoto, N. Yamazaki, and K. Yonezawa. 1979. Free-radical ring-opening polymerization. *Contemporary Topics in Polymer Science* 3:29–53.

Clendinning, R. A., S. Cohen, and J. E. Pott. 1974 (recd. 1975). Biodegradable containers — degradation rates and fabrication techniques. *Great Plains Agricultural Council Publication* 68(recd. 1975):244–54.

Doi, Y., Y. Kanesawa, M. Kunioka, and T. Saito, 1990. Biodegradation of microbial copolyesters: poly(3-hydroxybutyrate-co-3-hydroxyvalerate) and poly(3-hydroxybutyrate-co-4-hydroxybutyrate). *Macromolecules* 23:26–31.

Griffin, G. J. L. 1977. Degradation of polyethylene in compost burial. *Journal of Polymer Science* 57:281–86.

Gu, J. D., S. P. McCarthy, G. P. Smith, D. Eberiel, and R. Gross, 1992. Degradability of cellulose acetate (1.7 DS (degree of substitution)) and cellophane in anaerobic bioreactors. *Polymeric Materials Science and Engineering* 67:230–31.

Guillet, J. E. and Y. Americk. 1971. Photochemistry of ketone polymers. IV: Photolysis of methyl vinyl ketone copolymers. *Macromolecules* 4:375–79.

Hamilton, J. D., K. H. Reinert, J. V. Hagan, and W. V. Lord. 1995. Polymers as solid waste in municipal landfills. *Journal of the Air and Waste Management Association* 45:247–51.

Hamilton, J. D., M. B. Freeman, and K. H. Reinert. 1996. Aquatic risk assessment of a polycarboxylate dispersant used in laundry detergents. *Journal of Toxicology and Environmental Health.*

Hartley, G. H. and J. E. Guillet. 1968. Photochemistry of ketone polymers. I. Studies of ethylene–carbon monoxide copolymers. *Macromolecules* 1:165–70.

Heskins, M. and J. E. Guillet. 1968. Mechanism of ultraviolet stabilization of polymers. *Macromolecules* 1:97–98.

Kawai, F. 1987. The biochemistry of degradation of polyethers. *CRC Critical Reviews of Biotechnology* 6:273–307.

Keeler, R. 1991. Don't let food go to waste — make plastic out of it. *R&D* 33:52–57.

Lenz, R. W. 1995. *JTEC Monograph on Biodegradable Polymers and Plastics in Japan: Research, Development, and Applications.* Baltimore, MD: Loyola College, Japanese Technology Evaluation Center.

Matsumura, S., M. Nishioka, H. Shigeno, T. Tanaka, and S. Yoshikawa. 1993. Builder performance in detergent formulations and biodegradability of partially dicarboxylated cellulose and amylose containing sugar residues in the backbone. *Angew. Makromol. Chem.* 205:117–29.

Meister, J. J., M. Chen, and J. Meng. 1991. Graft 1-phenylethylene copolymers of lignin. 1. Synthesis and proof of copolymerization. *Macromolecules* 24:6843–48.

Otey, F. H., R. P. Westhoff and W. M. Doane. 1987. Starch-based blown films. *Industrial Engineering and Chemical Research* 26:1659–63.

Sawada, H. 1994. Field testing of biodegradable plastics. *Biodegradable Plastics and Polymers* 12:298–312.

Scholtz, N. 1991. Coupling the OECD confirmatory test with ecotoxicity tests for determining the aquatic toxicity under field conditions. Communication 2: *Daphnia magna* multigeneration test. *Tenside, Surfactants, Detergents* 28:277–81.

Swisher, R. D. 1987. *Surfactant Biodegradation*, 2nd ed. New York: Marcel Dekker.

Chapter 16

Life-Cycle Assessment of Polymers

■ BRUCE W. VIGON, DUANE A. TOLLE, AND VINCENT MCGINNISS
Battelle—Columbus

INTRODUCTION

The life-cycle concept is a "cradle-to-grave" approach for analyzing environmental impacts of products, processes, or activities. This concept recognizes that all life-cycle stages, along with all steps of design (i.e., development), can have economic, environmental, and energy consumption consequences. An understanding of key relationships between life-cycle stages and development can assist one in understanding the effects of technology choices and help to eliminate adverse environmental consequences.

Life-Cycle Assessment (LCA) is an evolving, quantitative tool that incorporates the life-cycle concept into an analytical framework. The major benefits of an LCA are:

- It quantifies the energy use, material requirements, wastes, emissions, and potential environmental impacts associated with the entire life cycle of a product, process, or activity.
- It contributes to a quantitative understanding of the overall and interdependent nature of environmental consequences and human activities.
- It gives decision makers (e.g., those involved in research and development decisions) quantitative information on potential environmental impacts associated with the entire life cycle of a product, process, or activity; identifies environmental improvement opportunities; and highlights life-cycle stages or processes with significant data gaps.

LCA Methods

As originally defined by the Society of Environmental Toxicology and Chemistry (SETAC) (1991, 1993a, b), the three major components of an LCA, in sequential order, are Inventory Analysis (LCI), Impact Assessment (LCIA), and Improvement Assessment (LCImA). More recently, the third component of the LCA paradigm appears to have evolved to include interpretation. This broader process may include an LCImA.

LCA may be an iterative process, as the requirements for a subsequent component may necessitate additional data acquisition or more rigorous data analysis in one of the earlier components.

Life-Cycle Inventory Analysis (LCI)

The first component of an LCA is the inventory analysis (U.S. EPA, 1993a). LCI is a technical, data-based process of quantifying energy and raw material inputs, atmospheric emissions, waterborne emissions, solid wastes, and other releases from the entire life cycle of a product, package, process, material, or activity.

A goal definition and scoping step is essential in conducting a useful and acceptable LCA (SETAC, 1991, 1993a,b; U.S. EPA, 1993a). Depending on whether the goal is to conduct an LCI or a combined LCI and LCIA, the scope of the data collection may be expanded when both types of evaluations are planned. This step specifies the nature and the extent of data to be used and the requirements for clarity and completeness in analyzing and presenting the findings of the study. The scope and purpose statement typically explains in concise fashion why the study is being undertaken, who the intended audience may be, and what implications these factors might have for the manner in which the study is performed and reported. This portion of the study generally includes a description of the system (or systems in the case of a comparative analysis) to be studied, the time frame involved, and the overall level of study specificity. The last point is especially important because questions related to the ability of an inventory to support an impact assessment that will be based on exceedance of measurable thresholds are resolved at this point.

This goal definition and scoping step also should involve a clear statement of the functional unit, which is a measure of performance for the system. A comparison of any systems in an LCI is based on equivalent performance, and the functional unit provides a basis for translating the measure of performance into the physical material and energy quantities accounted for in the inventory. This unit has to be clearly defined, measurable, and applicable to both input and output data. Examples of functional units might be "number of square meters of surface covered by 1,000 gallons of paint" or "quantity of materials required to install stretch-in residential nylon carpeting in a standard-sized room (i.e., 300 ft^2)."

LCI makes extensive use of both emission and functional unit databases (e.g., U.S. EPA, 1995a). Databases and software that incorporate emission and functional unit information are described below (under "Software and Databases for Polymer LCA").

Life-Cycle Impact Assessment (LCIA)

LCIA is a technical, quantitative, and/or qualitative process used to characterize and assess the effects of the environmental burdens identified in the LCI (SETAC, 1993a). At the 1992 SETAC meeting in Sandestin, Florida, the phases in an LCIA were defined as classification, characterization, and valuation (SETAC, 1993b). More recent SETAC meetings, however, have added a normalization phase after characterization and have moved the valuation phase to a new LCA component called interpretation, which are both discussed below.

As discussed previously, the goal definition and scoping step for an LCA involves an evaluation of boundaries, assumptions, and limitations. Typically, this step will already have been conducted as part of an LCI for the same product or process. Thus, the results of the LCI should be evaluated to determine if additional inventory information is needed for a thorough LCIA. Topics that should be addressed during the LCIA scoping meeting include:

- Matching the LCIA scope to the scope for the LCI.
- Identifying potential impacts and data needs required for their evaluation.
- Identifying impact analysis data that are missing from the LCI or need quantification.
- Considering the reasons for conducting an LCIA and how the results will be used.
- Justifying exclusion of any elements.
- Determining the level of impacts to be evaluated (source, media, or receptor).
- Considering the audience to which results will be presented.
- Defining spatial and temporal boundaries.

Classification is the process of assignment and initial aggregation of LCI data to relatively homogeneous stressor categories (e.g., greenhouse gases or ozone depletion compounds) within three primary impact categories (i.e., human health, ecological health, and resource depletion) (SETAC, 1993a, b). This grouping may involve inclusion of one inventory entry in the LCI in more than one category (e.g., NO_x both acidification and eutrophication potential). Stressor-impact chains are used to link inventory chemicals or resource use with primary (initial) impacts, and these primary impacts may lead to secondary, tertiary, etc., impacts in the same or other environmental compartments. In order to evaluate the potential impact of a stressor, it is important to know its physical and chemical properties, its potential transformation products in different media, and potential routes of exposure. Potential lists of impacts associated with each of the three primary impact categories are provided by SETAC and other sources (Heijungs, 1992; SETAC, 1993a,b). Lists of chemicals associated with specific stressor categories are given in Heijungs (1992). SETAC (1993b) included a fourth primary impact category (i.e., indirect environmental impacts arising from direct impacts on social welfare), but most LCIAs exclude this category.

Characterization in LCIA is the analysis and estimation of the magnitudes of potential impacts on the three primary impact categories by each of the stressor categories through the application of impact assessment tools (SETAC, 1993b). The choice of impact assessment tool depends on the level of impact analysis and the objectives and scope of the LCIA. The following types of analysis, in order of increasing complexity, have been proposed by SETAC:

- *Loading Assessment:* data from the inventory are listed and frequently grouped according to their potential effects; data for stressors that belong to a particular impact category may be summed. The assumption is that having less of a stressor is best, without regard for the fact that some stressors have a much greater environmental impact than others.
- *Equivalency Assessment:* data are aggregated according to equivalency factors for individual impacts (e.g., ozone-depletion potential or acidification potential). The assumption is that having less of the chemicals with the greatest impact potential is better.
- *Toxicity, Persistence, and Bioaccumulation Potential:* data are grouped on the basis of physical, chemical, and toxicological properties of chemicals that determine exposure and type of effect. The assumption is that having less of the chemicals with the greatest impact potential is better.
- *Generic Exposure/Effects Assessment:* impacts are predicted/modeled on the basis of generic environmental information usually associated with a region or an ecosystem.

- *Site-Specific Exposure/Effects Assessment:* impacts are predicted/ modeled on the basis of environmental information specific to a particular site.

Normalization has been proposed as the third step in an LCIA by researchers in both the United States and Europe, to put characterization data for different impact categories into perspective prior to their valuation (Goedkoop, 1995; Guinée, 1995; SETAC, 1993a). This step includes consideration of the magnitude and the frequency of the effects for each impact category relative to the total impact in a specified geographic region. Depending on the nature of the impact, the geographic area involved in determining the normalization value could be global (e.g., ozone depletion), regional (e.g., acid rain), or local (e.g., inhalation toxicity).

Life-Cycle Improvement Assessment (LCImA)

An improvement assessment (LCImA) has been defined as that portion of an LCA in which the options for reducing the environmental impacts or burdens associated with energy and raw material use and waste emissions throughout the life cycle of a product, process, or activity are identified and evaluated (SETAC, 1993a; U.S. EPA, 1993a). Although a standardized framework for improvement assessment has not been agreed upon, there is a general understanding that it deals with identification and evaluation of options for environmental improvements in products and processes. This analysis may include both quantitative and qualitative measures of improvements. The improvement options selected can involve changes in the product (e.g., substituted ingredients or eliminated components) or changes in the way the product is used to perform its intended function.

A series of questions should be answered to assist in the selection of improvement options. Depending on the purpose of the LCA and the nature of the sponsor (e.g., single company or government agency), these questions can include general questions about the need/purpose of the product or the process and specific questions about minor modifications in the existing product or process. Examples of general questions about the need/purpose of the product or the process include (Ervin, 1991):

- Is the product/service serving an essential purpose?
- Can the characteristics of complementary products or services be changed to eliminate/reduce the need for this product or service?
- Can the functions of more than one product or service be combined?
- Could fewer models or styles serve the purpose?
- Can the life span of the product be increased?

Examples of specific questions about minor modifications in the existing product or process include:

- Are there areas where substitute materials or processes could conserve energy and/or materials?
- Can the weight, size, or volume of the product or its associated packaging be reduced?
- Can the product be concentrated or produced in bulk or larger sizes?
- Can toxic substances in the product or its associated manufacture, use, or disposal be eliminated or reduced?
- Can the product's repairability, recyclability, or reusability be increased?
- Can the overall product efficiency be increased?

Interpretation

Although there is still debate on the subject, the third component of the LCA paradigm appears to have evolved from an "improvement assessment" (SETAC, 1993a) to an "interpretation." This new terminology permits the scope of an interpretation effort to be directed at improvement applications, but it expands the considerations to go beyond a strictly environmental nature. Thus, interpretation can include a consideration of cost and performance. Furthermore, the valuation process, which SETAC originally defined as the third step in impact assessment (SETAC, 1993b), has been moved to become part of the interpretation component. The valuation step involves the assignment of relative values or weights to different impacts in order to integrate all impacts in a form that will allow decision makers to consider the full range of impacts across all categories. Because the valuation process is inherently subjective, a variety of decision theory techniques are typically used to make the process more rational.

Decision theory techniques provide an additional level of interpretation in which individual impact categories are weighted for comparison. These methods use both expert judgment and input from interested and/or affected parties. Three possible valuation methods described by SETAC (1993b) are: Multi-Attribute Utility Theory (MAUT), Analytic Hierarchy Process (AHP), and Impact Analysis Matrix (IAM). MAUT involves breaking down a complex problem, such as an LCA, into individual objectives (e.g., minimize energy use or minimize air pollution) and the attributes of each objective (e.g., energy use processes or sources of air pollution). Then, decision analysis is used to measure the degree to which an objective is achieved by management options affecting the levels of attributes. AHP is a recognized methodology for supporting decisions based on relative preferences (importance) of pertinent factors (Saaty, 1990). The AHP process involves a structured description of the hierarchical relationships among the problem elements, beginning with an overall goal statement and working down the branches of the tree through the major and minor decision criteria. Preferences are expressed by the group in a pairwise manner supported by a software package known as Expert Choice (EC). The IAM method is a qualitative, expert-judgment-based approach that directly uses the results of the LCA. It was developed as part of a broader assessment for evaluating the source reduction potential of halogenated solvents, which included the assessment of alternatives. This method allows direct comparison of the relative environmental burdens between different alternatives. Other methods based on multiattribute decision analysis (MADA) are potentially useful for LCA as well.

The rigor of the interpretation assessment component, including the decision to interpret the LCI data alone or to include an LCIA for evaluating these data, depends on the purpose for conducting the study. Generally, the intensity of the scientific rigor should be increased, and an LCIA should be added when the purpose for the LCA expands beyond internal company use for design considerations to include broad-based environmental claims to the general public.

LCA APPLICATIONS FOR POLYMERS

Life-cycle studies of polymer-based systems have been conducted since the early days of LCA. An example LCA model for a plastic bottle manufacturing and disposal system is shown in Figure 16.1. The early studies tended to concentrate almost exclusively on packaging applications of high volume commodity polymers, including polyethylene (HDPE, LDPE, and LLDPE), polypropylene (PP), polystyrene (PS), polyethylene terphthalate (PET), and polyvinyl chloride (PVC). This situation is slowly beginning to change as LCA practitioners began applying the methods to a wider variety of materials and product systems. More recently, studies involving envineering polymers such as nylon, polycarbonate (PC), and rigid polyurethanes (rPU) have been conducted.

Because LCA has tended to focus on product systems, and more specifically on those systems intended to provide a consumer function, segregation of the contribution of the polymeric components to the overall life cycle is difficult. Furthermore, the majority of LCA studies have been performed for internal decision making in industrial materials and process selection. The resulting analyses have not been published in the open

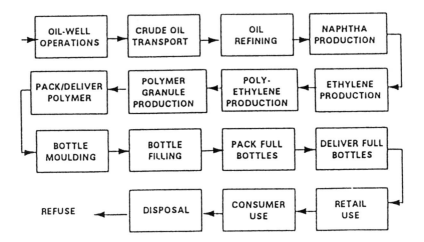

Figure 16.1. Main sequence of operations in the production and the use of a polyethylene bottle (SETAC, 1994).

literature and, except for limited presentations of research results at conferences and seminars, have not generally been discussed in the public forum.

As examples in the following section will show, the current picture of polymer LCA is both incomplete and intimately intertwined with the life cycles of other materials. The following sections describe LCA application to polymer design, environmental impact reduction, pollution prevention (P2), materials selection/product stewardship, and strategic product line management.

Polymer Design

General Aspects

When one designs a polymer system, it is important to consider the end use application, the expected performance requirements, and the best methods to produce the polymer with a given price/performance goal. Environmental factors now play an important role in the manufacture/production of the polymer, processing of the polymer into a product, and finally reclaiming or recycling the polymer product after is has served its useful purpose. For example, if the end product is to be reclaimed or eventually recycled after its useful lifetime, it probably should be made of a thermoplastic polymer, as such materials can be easily reclaimed, remelted, and made into similar or different types of products. Polyolefin milk bottles, polystyrene cups, and polyester bottles are a few of the common thermoplastic polymers in use today that can be easily recycled. In some cases, plastic products can be chemically or thermally degraded back into the polymer's original monomers or building blocks (e.g., base hydrolysis of polyesters into glycols and dibasic acids; thermal pyrolysis of polyethylene to produce ethylene).

Plastics

Performance aspects of polymers are broadly based on their designation as either thermoplastic or thermosetting materials. Thermoplastic products can be remelted or recycled and used again for a variety of applications; they are impact-resistant, but tend to have lower chemical resistance and lower continuous thermal service use properties than thermosetting polymer structures. Designed to be used in more demanding end use applications than thermoplastic polymers, thermosetting polymers include epoxies, phenolics, and unsaturated polyesters; they have high heat resistance, do not reflow under thermal stress, resist creep, and resist attack from a variety of chemicals and solvents. Examples of some common thermoplastic polymers are acetals, acrylic sheets, nylon 6, phenylene sulfide, polyester, PE, PP, PC, PS, and PVC.

In recent years, there has been some concern about the release of potentially harmful additives or chemicals during the processing of thermoplastics. With the assistance of LCA information, this concern has resulted in better design of plastics through the use of alternative monomers and/or catalyst systems to produce polymeric materials. Better purification procedures and improved polymer stabilization additives also have been developed so that the processing of polymers (e.g., molding, extruding) will no longer release harmful chemicals into the workplace or the final product during its expected lifetime.

Thermosetting polymers are much more difficult to recycle than their thermoplastic polymer counterparts, but LCA has shown that they can be effectively reground and used as fillers in other types of polymer processing operations. Some of the thermosetting polymers also can be chemically degraded into their individual monomers and used in the production of other polymer systems.

Packaging and Consumer Product/Engineering Polymers

This category of polymers has by far the largest amount of published life-cycle data. Comprehensive North American plastics industry inventory data are available for PE (high and low density), PP, PS, PET, and PVC (B. Vignon, Battelle, and K. McBride, American Plastics Council, personal communication, 1996). These data will be published as aggregate summaries for the raw materials extraction through resin production stages (i.e., cradle-to-gate). Details on resource, energy, and emissions may be obtained by contacting APC.

Life-cycle studies of products incorporating commodity and engineering polymers are widely available. Polymer materials covered include PS, PE, PP, PVC, PET, PC, nylon, and polyurethane foams. Comparison analyses involving various types of polymers are the most common, especially comparisons of lightweight polymer-based containers relative to glass, steel, or aluminum versions. Examples of published packaging studies in North America during the past decade include those concerned with beverage containers, drink cups, grocery bags, and fast food packaging (Council for Solid Waste Solutions, 1990a, b; Environmental Defense Fund, 1990; Hocking, 1991).

Nylon has been included as a component in two studies related to floor coverings and one related to automotive parts (Potting and Blok, 1995; Schuckert et al., 1993; U.S. EPA, 1993b). In the following example, the environmental loadings associated with a tufted carpet with a nylon pile, styrene-butadiene resin (SBR), and polypropylene backing were assessed (Table 16.1). The environmental aspects of carpet laying and subsequent cleaning for the 15-year life of the carpet were not included. About 65% ($154 \, MJ/m^2$) of the total use of primary energy in the floor covering system is associated with the energy inherent in nylon, and another 20% is associated with the energy inherent in SBR. Emissions to air of carbon dioxide (CO_2), carbon monoxide (CO), and sulfur dioxide (SO_2) are primarily associated with the need to generate energy for manufacturing processes. Emissions of styrene, ethylene, and tetrachloroethylene are associated with the manufacturing process itself, as are most of the waterborne emissions. Solid waste is primarily associated with post-consumer disposal of the used carpet, but about 20% of the waste is manufacturing-related.

In the automotive parts study, a comparison was made between an intake manifold manufactured from the traditional aluminum alloy and one made from nylon 6,6. No recycling of the nylon manifold was assumed because of a lack of recycling infrastructure; instead the part was either disposed of as solid waste or incinerated for energy recovery. A summary of the results is provided in Table 16.2. The data shown include not only the manufacturing, use, and ultimate disposition of the part itself, but also the effect of the part on the life-cycle energy efficiency of the vehicle.

Water-Soluble Polymers

Water-soluble polymers include a broad range of materials with molecular weights typically of several hundred thousand or lower. Some of the

Table 16.1. Life-cycle inventory summary for nylon carpet[a]

Parameter	Inventory results, MJ or g/FU[b]
Category: Energy Consumption, MJ	
Energy	154.3
Category: Emissions to Air, g	
Particulates	163.4
Hydrocarbons	31.0
Carbon dioxide (CO_2)	13,400
Carbon monoxide (CO)	16.1
Oxides of nitrogen (NO_x)	106.7
Sulfur dioxide (SO_2)	6.3
Tetrachloroethylene	0.3
Styrene	0.1
Ethylene	0.005
Aromatic hydrocarbons	6.8
Nitrous oxide (N_2O)	0.4
Ammonia	0.0002
Category: Emissions to Water, g	
Dyestuffs	2.4
Suspended particulates	0.002
Soluble inorganics	87.1
Soluble organics	19.9
Oil	0.8
Phenol	0.02
Fluorides	0.002
Category: Solid/Hazardous Waste, g	
Hazardous waste	645
Nonhazardous waste	2,770

[a]*Source:* Potting and Blok, 1995.

[b]Functional Unit (FU) = 1 m² of installed carpet with a lifetime of 15 years.

polymers, however, have molecular weights ranging up to about 10 million, having a polar or an amphoteric character that allows their use as soluble constituents in aqueous systems. Water-soluble polymers include polyethylene and polypropylene oxides (PEO/PPO), polyethylene and polypropylene glycols (PEG/PPG), polyvinyl alcohol (PVA), polyvinyl pyrollidone (PVP), polyethyleneimine (PEI), polyacrylamides (PA) polyacrylic acids and their corresponding salts (PAA), and various natural-based polymers such as polysaccharides (PSD), poly(lactic acid)s (PLA), and lignosulfonates (LS). Several of the water-soluble polymers (PA and PAA in particular) are routinely used as coagulants in water treatment and will be discussed below.

Polyacrylamides and polyacrylic acids (PA/PAA) are widely employed for the removal of fine particulate materials from water and wastewater. Although this is a major use of PA/PAA on a total annual quantity basis, life-cycle studies have not included these materials directly because they are used in amounts that are negligible in relation to the primary product of interest. Unless a product system used extraordinary amounts of treated water per functional unit of product, the quantity of polymer input to the system would fall below the 1% to 3% by mass that most practitioners use as a cutoff criterion on a formulated product basis.

Some other applications of PA/PAA include paper manufacturing and the production of superabsorbent gelling material for disposable diapers. Inventory profiles for the polymeric acrylate components of these products

Table 16.2. Life-cycle inventory summary of aluminum versus nylon intake manifold[a]

Parameter	Aluminum manifold, MJ or g/FU[b]	Nylon manifold, MJ or g/FU
Production energy	335	332
Use energy	1,021	408
Carbon dioxide (CO_2)	89.6	41.7
Dust	4.67	4.72
Carbon monoxide (CO)	17.1	9.1
Hydrocarbons	2.6	56.0
Nitrogen dioxide (NO_2)	39.7	51.4
Sulfur dioxide (SO_2)	57.7	42.3
Nitrous oxide (N_2O)	29.8	5.51
Hydrogen fluoride (HF)	0.21	0.02
Heavy metals	0.001	0.004
Hydrogen chloride (HCl)	1.97	0.013
Chlorine (Cl_2)	0.15	0.0

[a]*Source:* Schuckert et al., 1993.

[b]Functional Unit (FU) = 1 manifold of specified design installed in vehicle driven 200,000 km.

have been included in overall systems analysis. Franklin Associates, Ltd. (1992) considered sodium polyacrylate production from caustic soda and acrylic acid. The acrylic acid, in turn, was derived from ethylene oxide (presumably via the ethylene cyanohydrin route although hydrogen cyanide production is not shown as part of their process flow diagram). Polymerization initiators/catalysts include *t*-butyl hydroperoxide, persulfates, chlorates, and other sources of free radicals. Photo- and radiation-initiated polymerization also is possible. As with acrylic acid, acrylamide may be synthesized in a variety of ways, but production from acrylonitrile via reaction with acid is the typical practice. Again, inventory information for these polymers has not been published independently of associated consumer product systems.

Antifreeze Polymers

Based on available published information, polyethylene/propylene oxide and polyethylene/propylene glycol have not been included in life-cycle studies as stand-alone items. The environmental profiles of the monomers, however, have been investigated as part of a number of LCI studies. For example, in a comparison of antifreeze formulations, the ethylene and propylene oxide/glycol life-cycle inventories were presented by Franklin Associates, Ltd. (1994). The LCI indicated that conversion of the monomer to the respective polymer is typically a relatively clean process because the polymerization initiators are consumed in the reaction, and the reaction itself is exothermic. Therefore, production of emissions due to process releases and the need to generate steam or electricity to drive the reaction is very small.

Paints and Coatings

A recently completed effort in the United States applied LCA methods to the identification of improvements in chemical agent resistant coatings (CARC) for military vehicles and other items (U.S. EPA, 1995b). A life-cycle inventory and an impact assessment were carried out to compare the baseline system to five alternatives consisting of various combinations of alternative primer, alternative thinner, and alternative spray gun. One set of alternatives considered the simultaneous substitution of the alternative primer and gun. In addition to environmental considerations, the improvement assessment incorporated performance and cost aspects as a formal, structured part of the decision-making process.

The baseline CARC topcoat was a one-part, solvent-thinned polyurethane material. No substitutes for the topcoat were investigated, but an alternative thinner was evaluated. The baseline primer was a two-part, solvent-thinned epoxy, and the alternative was a two-part water-thinned epoxy. Application of impact assessment equivalency criteria and incorporation of cost/performance factors indicated that the alternative primer and the primer/gun combination offered the greatest life-cycle improvement potential.

Elastomers and Adhesives

Life-cycle data for adhesives and elastomers have not been published in any cohesive fashion, but may be found scattered throughout the life-cycle literature whenever products are encountered with adhesively bonded parts. Unfortunately, the elastomers and adhesives portions of the data are not separated in these studies from the remainder of the product system. Styrene-butadiene rubber (SBR), various epoxies, high impact polystyrene (HIPS), and other elastomer-modified materials are all included in this category. SBR latex has been included in studies of residential carpeting as part of the backing system, as discussed above. Epoxies are discussed above in the paints category.

HIPS is included in the North American APC/EPIC database (B. Vignon, Battelle, and K. McBride, American Plastics Council, personal communication, 1996). Details can be obtained by contacting APC. The system examined PS modified by the addition of polybutadiene rubber (PBR).

Environmental Impact Reduction

Because of the comprehensive nature of an LCA, all of the environmental impact categories relevant to LCAs in general are potentially applicable to the design of polymers. These impact categories can be classified under the three general protection areas: human health impacts, ecological health impacts, and resource depletion impacts. Many of the individual impact areas relevant to human health are also relevant to ecological health (e.g., global warming, ozone depletion, toxicity, photochemical oxidant formation, and acidification). Resource depletion involves both abiotic (e.g., mineral and energy resources) and biotic (e.g., biodiversity) resources. The impact categories considered to be of primary concern in an LCA of CARC paints were as follows:

- Photochemical oxidant creation potential (smog formation potential; the chemical species involved in total volatile organic carbon (VOC) emissions are important because not all VOCs have the same smog formation potential).

- Ozone depletion potential (stratospheric ozone depletion).
- Acidification potential (acid rain/fog).
- Global warming potential (greenhouse effect potential).
- Human health inhalation toxicity (e.g., acute inhalation toxicity).
- Terrestrial toxicity (e.g., acute oral wildlife toxicity).
- Aquatic toxicity (e.g., acute fish toxicity).
- Land use (e.g., for solid waste disposal).
- Natural resource depletion (e.g., fossil fuels and minerals).

Many of these impact categories were also identified in a useful document on environmental improvements in paints and adhesives by Inform, Inc. (Young et al., 1994). The Inform study was conducted with the cooperation of major paint manufacturing companies. These environmental impact areas are expected to be of primary concern for life-cycle design of most types of polymers, including paints.

Pollution Prevention (P2) Programs

LCA can also assist in decision making about P2 activities from several different perspectives. Most important, LCA expands P2 considerations beyond their effectiveness in source reduction at the point of application (i.e., facility-specific emissions) to a more global level. For example, a P2 alternative may make a minor reduction in certain moderately harmful emissions during product fabrication but cause major increases in other extremely harmful emissions during other life-cycle stages (e.g., raw material extraction or intermediate material manufacture). LCA helps to identify these types of environmental impact tradeoffs between alternatives so that the P2 alternative resulting in the smallest total life-cycle impacts can be identified.

A simplified P2 factors method for comparing P2 alternatives from an LCA perspective has been described by U.S. EPA (1994). This is a screening-level tool that focuses on only the LCA stages where differences in environmental impacts between two P2 alternatives are expected, based on stressor/impact diagrams. However, one can use the P2 factor score along with other factors such as cost, manufacturing potential, and performance to choose among similar P2 alternatives.

LCA helps to identify which processes in the life cycle of a product have the greatest opportunity for environmental impact reduction. This encourages development of P2 alternatives in the areas with the greatest potential for environmental improvement.

Materials Selection/Product Stewardship

The purpose of product stewardship is to make health, safety, and environmental protection an integral part of the product life cycle from design to disposal. LCA can help manufacturers during product design to select the materials for new or redesigned products that have the lowest total life-cycle environmental impact. The product stewardship concept involves consideration of potential impacts that may not be under the direct control of the product manufacturer. For materials that do not have reasonable substitutes, manufacturers can use LCA to select suppliers using intermediate material manufacturing methods that are environmentally superior. LCA can also form the basis for a corporation to select environmental stewardship goals that go beyond minimum legal compliance but are still economically favorable.

Strategic Product Line Management

Strategic product line management involves the use of a screening matrix to evaluate which products of a particular company need improvement from a competitive standpoint and/or a life-cycle environmental vulnerability standpoint. The conceptual model shown in Figure 16.2 is a two-dimensional graph with environmental vulnerability on the vertical axis and a measure of the product's competitive advantage on the horizontal axis. For the environmental vulnerability index, items in a set of qualitative environmental criteria (e.g., human toxicity, ozone depletion, global warming) are scored on a relative basis for each stage of the product life cycle. A similar method is used to score the product's competitive advantage, based on information from the company's business and marketing managers. Products that score well, with both a high competitive advantage and very low environmental vulnerability (lower right quadrant), can remain relatively unchanged. On the other hand, products that have a low competitive advantage and high environmental vulnerability (upper left quadrant) need substantial improvement in both areas, or strong consideration should be given to discontinuing them.

SOFTWARE AND DATABASES FOR POLYMER LCA

The conduct of polymer LCA requires familiarity with the availability and the use of existing software and polymer databases, as well as an understanding of the LCA methods that were outlined above.

Product Strategy

Figure 16.2. Strategic product line management and LCA concepts.

Process Flow Charts

The best way to understand and represent the components of a system is to produce a box and arrow drawing showing how each subsystem is inter-linked with others (SETAC, 1993a). Preparation of an LCA "model" can range from an oversimplified linear sequence of boxes showing only the main production processes to a detailed system flow diagram with intercon-necting "subsystems." Most industrial systems consist of the following three groups of operations: the main production sequence, the production of ancillary materials, and fuel production.

Databases: General Aspects

A preferred source of LCA data is data from companies operating the specific process (SETAC, 1993a; U.S. EPA, 1993a). Data are often available in several categories, depending on their state of aggregation. Besides influencing users' ability to interpret the inventory itself, highly aggregated data are not very helpful for impact assessment. Typical data categories are:

- *Individual process and facility-specific:* data of the most disaggregated type, which when coupled with the least amount of parameter aggregation (e.g., grouping of individual pollutants under a category name), provide the highest interpretation potential.

- *Composite:* data from the same operation combined across locations (e.g., the average propylene oxide production process for locations owned by a single company).
- *Aggregated or segmented:* data combining more than one process operation (e.g., polyvinyl chloride production via the suspension polymerization process from natural gas).
- *Industry-average:* data derived from a representative sample of locations and believed to statistically describe the typical operation across technologies (e.g., average U.S. national production of polyethylene from ethylene).
- *Generic:* data whose representativeness may be unknown but which are qualitatively descriptive of a process or a technology.

Data categories where location or other information desirable for estimating impact may have been lost through the averaging process are not well suited for inclusion in an impact assessment to be conducted where threshold exceedance will be measured. Unfortunately, other than for data sets derived directly from sponsoring organization audits or surveys, individual process and facility–specific data sets are the exception. In many instances, industry-supported studies collect data at the facility level, but for antitrust and business sensitivity reasons these data are reported at the aggregated or segmented level.

When primary data sources are not available, secondary data sources might include:

- Process designers.
- Engineering calculations based on process chemistry and technology.
- Estimations from similar operations.
- Published sources and commercial available databases.
- Marketplace patterns of usage of products.

Inventory checklists, such as those of U.S. EPA (1993a), are helpful in guiding data collection and simplifying construction of a computational model.

The quality of the data used in an LCI will significantly affect the results; so the development of uniform criteria is crucial for selection and reporting of data sources and types (U.S. EPA, 1993a; SETAC, 1994). Considerations should include data age, frequency of data collection, and representativeness of the data. Traditional indices of data quality that need to be considered include accuracy, precision, detection limits, and completeness.

Stand-alone data is a term sometimes used to describe the set of information developed to normalize individual subsystem inputs and outputs for the specific product or activity being analyzed. Each subsystem is reported in terms of a standard quantity of output, and then the output of each subsystem is adjusted to that needed for the overall product (i.e., use of the functional unit discussed earlier).

The next step is LCA model construction, consisting of incorporating the data and materials flows, as described in the system diagram, into a computational framework. The framework may be supported by a simple spreadsheet or sophisticated commercial LCA software. In either event, the framework not only combines the subsystems in a manner dictated by the flow diagram, functional unit, and coproduct allocation scheme; it also facilitates units conversion, energy resource acquisition factor analysis, sensitivity analysis, and data quality assessment. Most, but not all, software for performing inventory analysis is separate and distinct from the impact assessment modules or programs. Thus, the influence of one type of software on the other is minimal.

Example Databases

The conduct of LCAs is a data-intensive endeavor. The use of electronic databases, either public or private, can markedly increase the efficiency of performing an LCA. In many databases the topic of polymers is specifically limited to commodity materials primarily used in packaging. However, because some databases include segments that represent multiple processes, additional data on polymers may be embedded in these data sets. Users wishing to attempt separation of the polymer data should contact the database vendor to discuss their needs. A list of stand-alone databases is provided in Table 16.3; however, many LCA computer software packages also contain databases.

Databases may be categorized as follows:

- *Nonbibliographic databases*, containing on-line information covering resource use, energy consumption, environmental emissions, and chemical, biological, or toxicological effects.
- *Database clearinghouses*, which are on-line servics, both government-run and private, that facilitate the retrieval of information from a variety of databases and bibliographies.
- *Bibliographic databases*, which are on-line databases containing bibliographic references to the actual data, which must be extracted from the referenced sources.

Table 16.3. Stand-alone LCA databases

Database name	Author or organization	Address	City/county
North American Plastics Industry	American Plastics Council	1275 K Street, N.W., Suite 400	Washington, DC 20005
BUWAL	ETH, Institute for Energy and Cryoprocesses	ETH Zentrum	CH-8092, Zurich, Switzerland
Canadian Raw Materials Database	Environment Canada, Solid Waste Management Division	PVM 12	Ottawa, Ontario K1A 0H3
EMPA	Swiss Federal Laboratories for Materials Testing and Research	Unterstrasse 11	Ch-9001 St. Gallen, Switzerland
Environmental Resource Guide	American Institute of Architects, ERG Project	1735 New York Avenue, NW	Washington DC 20006
Plastic Waste Management Institute	European Centre for Plastics in the Environment	Avenue E. Van Nieuwenhuyse 4, Box 5	B-1160, Brussels, Belgium

Databases that are considered stand-alone are available as distinct information products separate from LCA software systems. Very few databases sold with LCA software packages are available as stand-alone products. Although a significant fraction of LCA data comes from product, material production (cradle-to-gate), or study-specific data collection efforts, the remainder is derived from so-called secondary data sources. Secondary data are defined as publicly available data that have not been collected specifically for the purpose of conducting LCAs, and for which the LCA practitioner has no input into the data collection process (Ryding et al., 1994).

A number of databases have been and continue to be developed specifically for LCA. They are sometimes, but not always, available independently of the computational software with which they are associated. In North America several notable efforts either have been completed or are under way to provide stand-alone, public database support for LCA:

- *Life-Cycle Inventory (Phase 1: Cradle-to-Pellet) of the North American Plastics Industry*, which is sponsored by the American Plastics Council and the Environment and Plastics Institute of Canada. It provides peer-reviewed, summary-level inventory data for five commodity polymers (polyethylene, polypropylene, polystyrene, polyvinyl chloride, and polyethylene terphthalate), three engineering polymers (acrylonitrile butadiene styrene copolymer, polycarbonate, and nylon, both 6 and 6/6 forms), and polyurethane feedstocks (methyldiisocyanate, toluene diisocyanate, and various polyols). Future phases will provide data on other life-cycle stages. The data represent primary information compiled from a representative set of producers for the polymers and their primary feedstock producers. For upstream products or processes (primarily petroleum/natural gas processing and petrochemical operations) and raw materials, data were derived from public sources and practitioner data cross-checked for accuracy and reasonableness. Data quality indicators were applied to the various data sets, weighted by the production of the individua producer. The highest quality was assigned to the measured, unallocated data and the lowest to estimated information.

- *Canadian Raw Materials Database (CRMD)*. The CRMD provides a peer-reviewed and arbitrated set of inventory data for a selected set of commonly used materials including various thermoplastics as well as nonplastic materials such as paper, aluminum, steel, glass, and wood. The data are collected and processed according to a rigorous, consensus-based procedure with the intent to publish a single, best set of data representing the national average. Quality metrics will be applied to screen data for inclusion in the database.

- *Environmental Resource Guide (ERG)*, sponsored by the American Institute of Architects and the U.S. Environmental Protection Agency. Although not an electronic database, the ERG contains a structured presentation of quantitative and qualitative information across the life cycle of various building and interior decorating materials. Materials reports include both polymeric and nonpolymeric materials: concrete, brick and mortar, concrete masonry, aluminum, steel, wood, plywood, particleboard, plastic laminates, thermal insulation, sealants, glass, plaster and lath, gypsum board, ceramic tile, ceiling panels and tiles, floor coverings (e.g., linoleum, vinyl, and carpet), paint, and wallcoverings.

Other North American database developments have occurred and are occurring in the context of LCA software commercialization. These include:

- *Life-Cycle Computer-Aided Data (LCAD)*, sponsored by the U.S. Department of Energy, Office of Industrial Technologies. LCAD provides data modules from raw material extraction to intermediate materials (cradle-to-gate) for major commodity materials including various commodity plastics (derived from the APC/EPIC database described above), commodity chemicals, paper, steel, aluminum, and electricity. The data are for U.S. average conditions except where a regional or an international perspective is needed to appropriately describe the operations. Data will be quality-assessed, with the characteristics included as part of the database. The database is intended for distribution as part of an LCA modeling software package.

- *Various private databases from North American LCA practitioners.* As part of the lease or purchase agreement for LCI software, several packages provide data sets as part of the arrangement. Typically, users can access the data themselves either not at all or only with some difficulty.

Several European LCA databases also exist, which may contain some pertinent information. In most cases these databases are not stand-alone entities but are associated with spreadsheets or other more sophisticated LCA models. They include:

- *BUWAL '84, '90, and '95.* The BUWAL (Swiss Office of Environmental Protection) database was originally developed to support Eco-profile Analysis within Switzerland. In 1990 and again in 1995 the Swiss Federal Technical Institute (ETH) updated and expanded the database. The stand-alone data are available in hardcopy and generally are applicable to Western European conditions although portions are limited to Switzerland. Most of the data pertain to packaging materials.
- *EMPA.* This database has been developed by the Swiss Federal Laboratory for Testing and Materials and is associated with an LCA model. In addition to energy, resource input, and environmental emissions data for a wide variety of processes, there is information on transportation and electricity production. In some cases the data are derived from secondary sources. The impact concept of critical volumes is incorporated in the system. Most of the data are specific to Swiss conditions, but the secondary data are more broadly applicable.
- *Packaging Industry Research Association (PIRA).* This database has been created by PIRA International with data derived from a study commissioned by the British government. The data are contained in an Excel spreadsheet and include a computational model for inventory analysis. Standard materials covered include virgin and recycled paper and board, steel tinplates, glass, and aluminum, as well as plastics data from the PWMI database listed below.
- *Plastic Waste Management Institute.* This database (Phase I) contains plastic resin information from raw materials acquisition through the

pellet stage for eight commodity thermoplastics. It is similar in content and approach to the APC/EPIC database for North America described above. Data represent European average and in some cases country-specific conditions.

Secondary data sources can be very useful in the conduct of LCAs when used with an appropriate understanding of their limitations. These sources include:

- Chemical technology reference books.
- Industry reports, such as SRI International studies on economic and production statistics.
- Open literature, including journal articles and patent summaries.
- U.S. Environment Protection Agency databases covering chemical emissions and disposition.
- U.S. Department of Energy databases, especially those of the Energy Information Administration.
- U.S. Department of the Interior data on natural resources.

As secondary data sources have been developed for purposes other than LCA, they are all likely to have substantial data gaps for any particular LCA (e.g., the lack of transportation energy information in nonbibliographic databases and the general lack of coverage of individual manufacturing processes). Prospective users must review data sources to determine which are most relevant for their purposes.

LCA Software

LCA has come of age in the era of personal computers. Although practitioners have used some form of computational support since the advent of life-cycle energy analysis studies in the 1970s, there has been an explosion of software tools in the 1990s. Software for LCA has permitted the extension and the implementation of a rigorous approach for both the analytical framework and the collection and the processing of data. As aptly noted by Heijungs and Guinée (1993): "The development of methodology for LCA is highly theoretical, whereas the collection of data has a direct connection with practice. The development of software increases the practical useability of the methodology and the suitability of the data within the theoretical framework. Software may thus act as a bridge between theory and practice." Table 16.4 lists examples of software available for LCA.

By requiring the user to supply certain information items, software can help to ensure that LCAs are conducted in full recognition of the

Table 16.4. Currently available and prototype LCA software packages

Model name	Author and organization	Address	City/county
Boustead	Dr. Ian Bousted	2 Black Cottages, West Grinstead, Horsham	West Sussex GB-RH-13 7BD Great Britain
EcoManager EcoManager	Franklin Associates Ltd. PIRA International	4121 W. 83rd Street, Suite 108 Randall Road, Surrey	Prairie Village, Kansas 66208 Leatherhead, KT22 7RU United Kingdom
ECOPACK 2000 EcoPro	Dipl. Chemiker Max, Bolliger EMPA (Swiss Federal Laboratories for Materials Testing and Research)	Esslenstrasse 26 Unterstrasse 11	CH-8280, Kreuzlingen, Switzerland Postfach CH-9001, St. Gallen, Switzerland
EcoSys	Sandia National Laboratories Laboratories	MS 0730	Albuquerque, New Mexico 87185-0730
EPRI Comprehensive Least Emissions ANalysis (CLEAN)	Science Applications International Corporation	5150 El Camino Real, Suite C-31	Los Altos, California 94022
Environmental Priorities Strategies System (EPS)	Swedish Environmental Research Institute (IVL)	Box 21060	S-10031 Stockholm, Sweden
International Database for Ecoprofile Analysis (IDEA)	International Institute for Applied Systems Analysis (IIASA)	1220 Wien, Bernoullistrasse	4/5/6/19 Laxenburg, Austria
KCL-ECO	Oy Keskuslaboratorio— Centrallaboratorium Ab, The Finnish Pulp and Paper Research Institute	Tekniikantie 2	FIN-02150 Espoo, Finland

Tool	Organization	Address	
LCI 1 Spreadsheet	Procter & Gamble, European Technical Center	Boechoutlaan 107	1820 Strombeek Bever, Belgium
LCA Inventory Tool (LCAiT)	PRé Consultants	Bergstraat 6	3811 NH Amersfoort, The Netherlands
LCA Inventory Tool (LCAiT)	Chalmers Industriteknik	Chalmers Teknikpark,	S-41288 Göteborg, Sweden
Life-Cycle Computer-Aided Data (LCAD)	Battelle Pacific Northwest Laboratories	Battelle Boulevard, P.O. Box 999	Richland, Washington 99352
Life-Cycle Interactive Modeling System (LiMS)	Chem Systems Inc.	303 S. Broadway	Tarrytown, New York 10591-5487
PIRA Environmental Management System (PEMS)	PIRA International	Randall Road, Surrey	Leatherhead, KT22 7RU United Kingdom
Product and Process Software	Flemish Institute for Technological Research (V.I.T.O.)	Boeretang 200	2400 Mol, Belgium
Product Improvement Assessment (PIA)	Institute for Applied Environmental Economics (TME)	Grote Markstraat 24	2511 BJ-Gravenhage, The Hague, The Netherlands
Resource and Environmental Profile Analysis Query (REPAQ)	Franklin Associates Ltd.	4121 W. 83rd Street, Suite 108	Prairie Village, Kansas 66208
SimaPro 3.1	PRé Consultants	Bergstraat 6	3811 NH Amersfoort, The Netherlands
Total Emission Model for Integrated Systems (TEMIS)	Öko-Institut (Institute for Applied Ecology), Energy Division, Central Office	Binzengrün 34 a	79114 Freiburg, Germany
Tools for Environmental Analysis and Management (TEAM)	Ecobalance Inc.	1 Church Street, Suite 700	Rockville, Maryland 20850

assumptions and the definitions involved. Software products typically comprise a user interface (graphics-based systems now challenge text-driven interfaces in commercial or near-commercial products), a database, a computational engine, and a report processor. The most common type of LCA model, as distinguished from design or engineering LCA modeling tools, is referred to as an input–output model. The intent of this type of model is to capture the materials and energy balance at the system or the subsystem level and not to focus on process details.

Product design–oriented LCA tools target a user audience that is not and probably will not become expert in LCA and has little or no knowledge or technical expertise in environmental issues. Product design software users can include mechanical design engineers, packaging designers, product concept specialists, and graphic designers whose primary interface with software is in computer-aided design (CAD) or mechanical/structural design packages. Product development support software incorporates LCA computations in a framework that has the appearance and the character of software within the domain of expertise of the designer. Product design–oriented software also includes software intended primarily for development of environmentally sensitive packaging.

LCA software products developed for these users typically result in a kind of "green advisor" that provides recommendations on materials and process choices based on life-cycle considerations. The user is prompted for information about the physical form and/or the layout of the components or the product and is given choices regarding materials that may be used in a given application. Some software also provides the ability to select alternative processes for manufacture of the item. Within the software, a database and an expert system have been incorporated to translate the designer's choices into the necessary inventory and impact assessment computations. Choices on the depth and the breadth of the LCA may have been preselected by the developer in order to balance the complexity with the multidimensional nature of the decision process (Wixom, 1994).

SUMMARY AND FUTURE DIRECTIONS

LCA is gaining acceptance in the development of clean polymer manufacturing processes and in the selection of polymeric materials for use in a variety of finished products. Life-cycle inventory analyses, in which resource consumption, energy utilization, and emissions to air, water, and land are tabulated, have been conducted for a number of polymers. Future polymer

LCA should bring even greater value and better integration of LCA with product support programs.

REFERENCES

Council for Solid Waste Solutions. 1990a. *Resource and Environmental Profile Analysis of Foam Polystyrene and Bleached Paperboard Containers.* Prairie Village, KS: Franklin Associates, Ltd.
Council for Solid Waste Solutions. 1990b. *Resource and Environmental Profile Analysis of Polyethylene and Unbleached Paper Grocery Sacks.* Franklin Associates, Ltd.
Environmental Defense Fund. 1990. *Good Things Come in Smaller Packages: The Technical and Economic Arguments in Support of McDonald's Decision to Phase Out Polystryrene Foam Packaging.* Washington, DC: Environmental Defense Fund.
Ervin, C. 1991. Source reduction: Getting beyond the rhetoric. *Pollution Prevention Review,* p. 115.
Franklin Associates, Ltd. 1992. Energy and environmental profile analysis of children's single use and cloth diapers, revised report. Prepared for American Paper Institute. Prairie Village, NY: Franklin Associates, Ltd.
Franklin Associates, Ltd. 1994. *Life-Cycle Assessment of Etylene and Propylene Glycol Antifreeze.* Prairie Village, NY: Franklin Associates, Ltd.
Goedkoop, M. 1995. The Eco-Indicator 95. Report 9523 for Novem (Netherlands Agency for Energy and the Environment) and RIVM (National Institute of Public Health and Environmental Protection). Amersfoort, The Netherlands: PRé Consultants.
Guinée, J. B. 1995. Development of a methodology for the environmental life-cycle assessment of products with a case study on margarines. Ph.D. thesis. Molenaarsgraaf, The Netherlands: Optima Druk Publishers.
Heijungs, R. (final ed.). 1992. Environmental life-cycle assessment of products: Guide — October 1992, report 9266. The Netherlands: CML (Centre of Environmental Science) in Leiden, TNO (Netherlands Organisation for Applied Scientific Research) in Apeldoorn, and B&G (Fuels and Raw Materials Bureau) in Rotterdam.
Heijungs, R. and J. Guinée. 1993. Software as a bridge between theory and practice in life-cycle assessments. *Journal of Cleaner Production* 1(3–4):185–92.
Hocking, M. B. 1991. Paper versus polystyrene: A complex choice. *Science* 25:504–5.

Potting, J. and K. Blok. 1995. Life-cycle assessment of floor covering. *Journal of Cleaner Production* 3:201–13.

Ryding, S-O., B. Steen, A. Wenblad, and R. Karlsson. 1994. The EPS system: A life-cycle assessment concept for cleaner technology and product development strategies, and design for the environment, in *Proceedings of the EPA Workshop on Identifying a Framework for Human Health and Environmental Risk Ranking,* June 30–July 1, 1993, EPA 744-S-93-001, pp. 21–24 and supplemental materials in Appendix G.

Saaty, T. L. 1990. *The Analytic Hierarchy Process.* Pittsburgh, PA: RWS Publications, 287 pp.

Schuckert, M., I. C. Pfleiderer, K. Saur, and P. Eyerer. 1993. Life-cycle analysis of automotive parts, interim report, in *Proceedings Eco-Balance Review Conference,* Brussels: Belgium.

Society of Environmental Toxicology and Chemistry (SETAC). 1991. *A Technical Framework for Life-cycle Assessments.* Pensacola, FL: Society of Environmental Toxicology and Chemistry and SETAC Foundation for Environmental Education, Inc.

Society of Environmental Toxicology and Chemistry (SETAC). 1993a. *Guidelines for Life-cycle Assessment: A Code of Practice.* Pensacola, FL: Society of Environmental Toxicology and Chemistry and SETAC Foundation for Environmental Education, Inc.

Society of Environmental Toxicology and Chemistry (SETAC). 1993b. *A Conceptual Framework for Life-Cycle Impact Assessment.* Pensacola, FL: Society of Environmental Toxicology and Chemistry and SETAC Foundation for Environmental Education, Inc.

Society of Environmental Toxicology and Chemistry (SETAC). 1994. *Life-Cycle Assessment Data Quality: A Conceptual Framework.* Pensacola, FL: Society of Environmental Toxicology and Chemistry and SETAC Foundation for Environmental Education, Inc.

U.S. Environmental Protection Agency (U.S. EPA). 1993a. *Life-Cycle Assessment: Inventory Guidelines and Principles.* EPA/600/R-92/245. Prepared by Battelle and Franklin Associates, Inc. Cincinnati, OH: Risk Reduction Engineering Laboratory, Office of Research and Development, U.S. Environmental Protection Agency.

U.S. Environmental Protection Agency (U.S. EPA). 1993b. *Life-Cycle Assessment Methodology: Inventory Demonstration Project — Residential Nylon Carpeting — Interim Report.* Prepared by Battelle. Cincinnati: OH: Risk Reduction Engineering Laboratory, Office of Research and Development, U.S. Environmental Protection Agency.

U.S. Environmental Protection Agency (U.S. EPA). 1994. *Development of a Pollution Prevention Factors Methodology Based on Life-Cycle Assessment: Lithographic Printing Case Study.* EPA/600/R-94/157. Prepared by Battelle. Cincinnati, OH: Risk Reduction Engineering Laboratory,

Office of Research and Development, U.S. Environmental Protection Agency.

U.S. Environmental Protection Agency (U.S. EPA). 1995a. *Life-Cycle Assessment: Public Data Sources for the LCA Practitioner.* EPA 530/R-095/009. Prepared by Battelle. Washington, DC: Office of Solid Waste.

U.S. Environmental Protection Agency (U.S. EPA). 1995b. *Life-Cycle Improvement Assessment (LCImA) for Chemical Agent Resistant Coating: Draft Final Report.* Prepared by Battelle. Washington, DC: National Risk Management Research Laboratory.

Wixom, M. R. 1994 The NCMS green design advisor: A CAE tool for environmentally conscious design, in *Proceedings International Symposium on Electronics and the Environment.* San Francisco, CA: Institute of Electrical and Electronic Engineers.

Young, J. S., L. Ambrose, and L. Lobo. 1994. *Stirring Up Innovation: Environmental Improvements in Paints and Adhesives.* New York: Inform, Inc.

Index

http://www.vnr.com

product discounts   free email newsletters
software demos   online resources

email: info@vnr.com   A service   I(T)P